Pollution Prevention through Process Integration

Pollution Prevention through Process Integration

Systematic Design Tools

Mahmoud M. El-Halwagi

Chemical Engineering Department

Auburn University

Auburn, Alabama

ACADEMIC PRESS

San Diego   London   Boston   New York   Sydney   Tokyo   Toronto

This book is printed on acid-free paper. ∞

Academic Press
a division of Harcourt Brace & Company
525 B Street, Suite 1900, San Diego, California 92101-4495, USA
http://www.apnet.com

Academic Press Limited
24-28 Oval Road, London NW1 7DX, UK
http://www.hbuk.co.uk/ap/

Library of Congress Cataloging-in-Publication Data

El-Halwagi, Mahmoud M., date.
    Pollution prevention through process integration : systematic design tools / by Mahmoud El-Halwagi.
            p.   cm.
    Includes index.
    ISBN 0-12-236845-2 (alk. paper)
    1. Chemical industries--Environmental aspects.   2. Pollution prevention.   3. Chemical process control.   I. Title.
TD195.C45E44   1997
660'.028'6--dc21   97-9789
                                                          CIP

PRINTED IN THE UNITED STATES OF AMERICA
97   98   99   00   01   02   MM   9   8   7   6   5   4   3   2   1

"And say O my Lord: Advance me in knowledge"
THE HOLY QURAN (XX:114)

To my parents, my wife, and my children with love.

Contents

CHAPTER THREE

Synthesis of Mass-Exchange Networks
A Graphical Approach

CHAPTER FOUR

Graphical Techniques for Mass Integration with Mass-Exchange Interception

CHAPTER FIVE

Synthesis of Mass-Exchange Networks—An Algebraic Approach

Preface

Processing facilities are complex systems of unit operations and streams. Consequently, their environmental impact cannot be optimally mitigated by simple end-of-pipe measures. Instead, it is crucial to gain global insights into how mass flows throughout the process and to use these insights as a consistent basis for developing cost-effective pollution-prevention solutions. These global insights can help in extracting simple solutions from complex processes without the need for laborious conventional engineering approaches.

Over the past decade, significant advances have been made in treating chemical processes as integrated systems. This holistic approach can be used to enhance and reconcile various process objectives, such as cost effectiveness, yield enhancement, energy efficiency, and pollution prevention. In this context, a particularly powerful concept is mass integration, which deals with the optimum routing of streams as well as allocation, generation, and separation of species. Many archival papers have been published on different aspects of mass integration. These papers have mostly targeted researchers in the field of process synthesis and design. This book was motivated by the need to reach out to a much wider base of readers who are interested in systematically addressing pollution prevention problems in a cost-effective manner.

This work is the first textbook of its kind, systematizing what is seemingly a pollution-prevention art that depends heavily on experience and subjective opinions into a science that is rooted in fundamental chemical engineering concepts and process integration. The book is intended to build a bridge between the academic world of fundamentals and the industrial world of applicability. It presents systematic and generally applicable techniques for cost-effective pollution prevention that are neither simple rules of thumb or heuristics nor all-inclusive sophisticated mathematical optimization programs geared exclusively toward academic researchers. Instead, these techniques are based on key scientific and engineering fundamentals and, therefore, provide a strong foundation for tackling a wide variety of environmental problems. They also provide different levels of sophistication and involvement ranging from graphical methods to algebraic procedures and mathematical optimization, thereby targeting a wide spectrum of practicing engineers and environmental professionals in the process industries and pollution prevention, upper-level undergraduate students and first-year graduate students, and researchers in the areas of process integration and pollution prevention. Graphical tools are useful in providing a clear visual representation of a system, and can be used to systematically guide the designer in generating solutions. However, algebraic procedures are computationally more convenient than graphical methods and can be easily implemented using calculators or spreadsheets. In principle, mathematical optimization is more powerful than other techniques but cannot easily incorporate the engineer's judgment and preference.

As a holistic approach to pollution prevention, the book addresses cost effectiveness and key technical objectives of processing facilities. In addition to the environmental objectives. A key philosophy of the book is first to establish a breadth of understanding and overall performance targets of the process using the global insights of process integration, and then to develop the necessary in-depth analysis. The importance of practical know-how in attaining the full potential of a new technology is recognized, so the applications presented in this book draw largely from the expertise of the author and his co-workers in applying mass integration technology to a wide variety of process industries.

I am grateful to colleagues, students and faculty at institutions where I learned and taught, including Auburn University, Cairo University and the University of California, Los Angeles (UCLA). I am thankful to Professor Vasilios Manousiothakis of UCLA, who helped me during my research beginnings in process synthesis and Professor Sheldon Friedlander, also of UCLA, who deeply influenced my vision on the next generation of pollution-prevention practices. I am indebted to the numerous undergraduate students at Auburn University and attendees of my industrial workshops who provided valuable comments on the notes that preceded this text. I am also grateful to the many researchers in the area of process synthesis and integration who have contributed to this emerging field over the past two decades.

The intellectual contributions and stimulating research of my former and current graduate students constitute key landmarks in preventing pollution via process integration. I had the distinct pleasure of working with and learning from this superb group of enthusiastic and self-motivated individuals. I am particularly indebted to Srinivas "B.K." Bagepalli (General Electric Corporate Research and Development), Eric Crabtree, Tony Davis (Lockheed Martin), Alec Dobson (Monsanto), Russell Dunn (Monsanto), Brent Ellison (Matrix Process Integration), Walker Garrison (Matrix Process Integration), Murali Gopalakrishnan, Ahmad Hamad, Eva Lovelady, Bahy Noureldin, Gautham "P.G." Pathasarathy, Andrea Richburg, (U.S. Department of Energy), Mark Shelley, Chris Soileau,

Carol Stanley (Linnhoff March), Obaid Yousuf (Platinum), Ragavan Vaidyanathan (M. W. Kellogg), Anthony Warren (General Electric Plastics), Matt Wolf (Allied Signal) and Mingjie Zhu (General Electric Plastics).

I also thank Dr. Dennis Spriggs, President of Matrix Process Integration, who injected into my research a valuable dimension of applicability and a deep appreciation for what it takes to develop a technology and transform it into industrial practice. I am grateful to him and to the engineering team of Matrix Process Integration for taking a pioneering role in applying mass integration to various industries, providing critical feedback, and developing valuable know-how.

I am indebted to various Federal and State agencies as well as companies for providing support to my research in pollution prevention through process integration. I am also grateful to the providers of awards that motivated me to transfer research accomplishments to the classroom including the National Science Foundations' National Young Investigator (NYI) Award, the Birdsong Merit Teaching Award, and the Fred H. Pumphrey Award.

I am thankful to LINDO Systems Inc. for providing the optimization software LINGO. I also appreciate the assistance provided by Mr. Brent Ellison and Mr. Obaid Yousuf in developing the MEN software.

I appreciate the fine work of the editing and production team at Academic Press. I am specially thankful to Dr. David Packer, Ms. Linda McAleer, Ms. Rebecca Orbegoso, and Ms. Jacqueline Garrett for their excellent work and cooperative spirit.

I am grateful to my mother for being a constant source of love and support throughout my life, and to my father, Dr. M. M. El-Halwagi, for being my most profound mentor, introducing me to the fascinating world of chemical and environmental engineering, and teaching me the most valuable notions in the profession and in life. I am also grateful to my grandfather, the late Dr. M. A. El-Halwagi, for instilling in me a deep love for chemical engineering and a desire for continued learning. I am indebted to my wife, Amal, for her personal and professional companionship over the past two decades. She has contributed many bright ideas to my research and to this book, and has always been my first reader and my most constructive critic. Without her great deal of love, support, encouragement, and never-ending patience, this work would not have been completed. Finally, I am grateful to my kids, Omar and Ali, for tolerating lost evenings, weekends and soccer games as "Dad was fighting the pollution in the air and the ocean."

Mahmoud M. El-Halwagi

CHAPTER ONE

Introduction

This chapter provides an overview of pollution prevention and process integration. First, the key strategies for reducing industrial waste are discussed. Then process integration is presented as a viable tool for systematizing pollution-prevention decisions. Three key elements of process integration are discussed: synthesis, analysis and optimization. Next, process integration is categorized into mass integration and energy integration, with special emphasis on mass integration as it plays a key role in pollution prevention. Finally, the scope and structure of the book are discussed.

1.1 The Environmental Problem and Pollution Prevention

Environmental impact is one of the most serious challenges currently facing the chemical process industry. In the United States alone, it is estimated that 12 billion tons (wet basis) of industrial waste are generated annually (Allen and Rosselot, 1994). The staggering magnitude of industrial waste coupled with the growing awareness of the consequences of discharging effluents into natural resources has spurred the process industry to become more environmentally conscious and adopt a more proactive role. Over the past two decades, significant efforts have been directed toward reducing industrial waste. The focus of these environmental efforts has gradually shifted from downstream pollution control to a more aggressive practice of trying to prevent pollution in the first place. In the 1970s, the main environmental activity of the process industries was end-of-the-pipe treatment. This approach is based on installing pollution control units that can reduce the load or toxicity of wastes to acceptable levels. Most of these units employ conversion techniques (e.g., incineration or biotreatment) that transform the contaminants into more benign species. In the 1980s, the chemical process industries have shown a strong interest in implementing recycle/reuse policies in which pollutants are recovered from terminal streams (typically using separation processes) and reused or sold. This approach has gained significant momentum from the realization that waste streams can be valuable process resources when tackled in a cost-effective manner. At present, there is a substantial industrial interest in the more comprehensive concept of pollution prevention.

Several definitions of pollution prevention can be found in the literature (e.g., El-Halwagi and Petrides, 1995; Freeman, 1995; Theodore *et al.*, 1994; Noyes, 1993). These definitions vary in the scope of pollution prevention. Throughout this book, the term pollution prevention will be used to describe any activity that is aimed at reducing, to the extent feasible, the release of undesirable substances to the environment. Other terms such as waste minimization, reduction, and management will be used interchangeably as synonyms for pollution prevention.

A hierarchy of four main strategies can be used to reduce the waste within a process. This hierarchy establishes the priority order in which waste management activities should be employed:

1. Source reduction includes any in-plant actions to reduce the quantity or the toxicity of the waste at the source. Examples include equipment modification, design and operational changes of the process, reformulation or redesign of products, substitution of raw materials, and use of environmentally benign chemical reactions.

2. Recycle/reuse involves the use of pollutant-laden streams within the process. Typically, separation technologies are key elements in a recycle/reuse system to recover valuable materials such as solvents, metals, inorganic species, and water.

3. End-of-pipe treatment refers to the application of chemical, biological, and physical processes to reduce the toxicity or volume of downstream waste. Treatment options include biological systems, chemical precipitation, flocculation, coagulation, and incineration as well as boilers and industrial furnaces (BIFs).

4. Disposal involves the use of postprocess activities that can handle waste, such as deep-well injection and off-site shipment of hazardous materials to waste-management facilities.

The focus of pollution prevention and this book is on the first two options: source reduction and recycle/reuse. Pollution can be more effectively handled by reducing upstream sources of pollutants than by using extensive downstream (or end-of-pipe) treatment processes. Equivalently, "Do your best, then treat the rest." Notwithstanding this focus, it is important to bear in mind that the four strategies should be integrated and reconciled. An effective design methodology must have the ability to determine the optimal extent to which each strategy should be used.

As a result of the growing interest in pollution prevention, several industries have been actively developing and implementing various strategies for pollution prevention. Most of these strategies have been tailored to solve specific problems for individual plants. This case-based approach is inherently limited because the pollution-prevention technology and expertise gained cannot be transferred to other plants and industries. The lack of generally applicable techniques renders the task of developing pollution-prevention strategies a laborious one. To develop pollution prevention solutions for a given industrial situation, engineers are typically confronted with design decisions that require making choices from a vast number of options. The engineers must select the type of pollution prevention technologies, system components, interconnection of units, and operating conditions. In many cases, there are too many alternatives to enumerate. These challenges call for the application of a systematic and generally applicable approach which transcends the specific circumstances of the process, views the environmental problem from a holistic perspective, and integrates:

- All process objectives including cost effectiveness, yield enhancement, energy conservation, and environmental acceptability.
- The waste receiving media namely air, water and land.
- The various waste-management options namely, source reduction, recycle/reuse, treatment, and disposal.
- All waste reduction technologies.

In this context, process integration can provide an excellent framework for addressing the foregoing objectives.

1.2 What Is Process Integration?

A chemical process is an integrated system of interconnected units and streams, and it should be treated as such. Process integration is a holistic approach to process design, retrofitting, and operation which emphasizes the unity of the process. In light of the strong interaction among process units, streams, and objectives, process integration offers a unique framework for fundamentally understanding the global insights of the process, methodically determining its attainable performance targets, and systematically making decisions leading to the realization of these targets. There are three key components in any comprehensive process integration methodology; synthesis, analysis, and optimization.

1.2.1 Process Synthesis

Process synthesis may be defined as (Westerberg, 1987): "the discrete decision-making activities of conjecturing (1) which of the many available component parts one should use, and (2) how they should be interconnected to structure the optimal solution to a given design problem." Therefore, the field of process synthesis is concerned with the activities in which the various process elements are integrated and the flowsheet of the system is generated so as to meet certain objectives. Process synthesis is a relatively new engineering discipline. Reviews of the field can be found in literature (e.g., El-Halwagi and El-Halwagi, 1992; Douglas, 1992; Westerberg, 1987; Stephanopoulos and Townsend, 1986; Nishida *et al.*, 1981).

Process synthesis provides an attractive framework for tackling numerous design problems through a systemic approach. It guides the designer in the generation and screening of various process technologies, alternatives, configurations, and operating conditions. In most applications, the number of process alternatives is too high (in many cases infinite). Without a systematic approach for process synthesis, an engineer normally synthesizes a few process alternatives based on experience and corporate preference. The designer then selects the alternative with the most promising economic potential and designates it as the "optimum" solution. However, by assessing only a limited number of alternatives one may easily miss the true optimum solution, or even become trapped in a region that is significantly different from the optimal one. In addition, the likelihood of generating innovative designs is severely reduced by an exclusive dependence on previous experience.

Because of the vast number of process alternatives, it is important that the synthesis techniques be able to extract the optimal solution(s) from among the numerous candidates without the need to enumerate these options. Two main synthesis approaches can be employed to determine solutions while circumventing the dimensionality problem; *structure independent and structure based.* The structure-independent (or targeting) approach is based on tackling the synthesis task via a sequence of stages. Within each stage, a design target can be identified and employed in subsequent stages. *Such targets are determined ahead of detailed design and without commitment to the final system configuration.* The targeting approach offers two main advantages. First, within each stage, the problem dimensionality is reduced to a manageable size, avoiding the combinatorial problems. Second, this approach offers valuable insights into the system performance and characteristics.

The second category of process-synthesis strategies is structural. This technique involves the development of a framework that embeds all potential configurations of interest. Examples of these frameworks include process graphs, state-space representations and superstructures (e.g., Friedler *et al.*, 1995; Bagajewicz and Manousiouthakis, 1992; Floudas *et al.*, 1986). The mathematical representation used in this approach is typically in the form of mixed-integer nonlinear programs, (MINLPs). The objective of these programs is to identify two types of variables; integer and continuous. The integer variables correspond to the existence or absence of certain technologies and pieces of equipment in the solution. For instance, a binary integer variable can assume a value of one when a unit is selected and zero when it is not chosen as part of the solution. On the other hand, the continuous variables determine the optimal values of nondiscrete design and operating parameters such as flowrates, temperatures, pressures, and unit sizes. Although this approach is potentially more robust than the structure-independent strategies, its success depends strongly on three challenging factors. First, the system representation should embed as many potential alternatives as possible. Failure to incorporate certain configurations may result in suboptimal solutions. Second, the nonlinearity properties of the mathematical formulations mean that obtaining a global solution to these optimization programs can sometimes be an illusive goal. This issue can be a major hurdle, as current commercial optimization software cannot guarantee the global solution of general MINLPs. Finally, once the synthesis task is formulated as an MINLP, the engineer's input, preference, judgment, and insights are set aside. Therefore, it is important to incorporate these insights as part of the problem formulation. This can be a tedious task.

The result of process synthesis is a flowsheet which represents the configuration of the various pieces of equipment and their interconnection. Next, it is necessary to analyze the performance of this flowsheet.

1.2.2 Process Analysis

While synthesis is aimed at combining the process elements into a coherent whole, analysis involves the decomposition of the whole into its constituent elements for individual study of performance. Hence, once a process is synthesized, its detailed characteristics (e.g., flowrates, compositions, temperature, and pressure) are predicted using analysis techniques. These techniques include mathematical models, empirical correlations, and computer-aided process simulation tools (e.g., ASPEN Plus, ChemCAD III, PRO II, HYSIM). In addition, process analysis may involve predicting and validating performance using experiments at the lab and pilot-plant scales, and even actual runs of existing facilities.

1.2.3 Process Optimization

Once the process has been synthesized and its performance has been characterized, one can determine whether or not the design objectives have been met. Therefore, synthesis and analysis activities are iteratively continued until the process objectives are realized. The realization of process objectives implies that we have a solution that works but not necessarily an optimum one. Therefore, it is necessary to include optimization in a comprehensive process integration methodology. Optimization involves the selection of the "best" solution from among the set of candidate solutions. The degree of goodness of the solution is quantified using an *objective function* (e.g., cost, profit, generated waste) which is to be minimized or maximized. The search process is undertaken subject to the system model and restrictions which are termed *constraints*. These constraints are in the form of equality and inequality expressions. Examples of equality constraints include material and energy balances, process modeling equations, and thermodynamic requirements. On the other hand, the nature of inequality constraints may be environmental (e.g., the quantity of certain pollutants should be below specific levels), technical (e.g., pressure, temperature or flowrate should not exceed some given values) or thermodynamic (e.g., the state of the system cannot violate second law of thermodynamics). An optimization problem in which the objective function as well as all the constraints are linear is called a linear program (LP); otherwise it is termed a nonlinear program (NLP). The nature of optimization variables also affects the classification of optimization programs. An optimization formulation that contains continuous (real) variables (e.g., pressure, temperature, or flowrate) as well as integer variables (e.g., 0, 1, 2, . . .) is called a mixed-integer program (MIP). Depending on the linearity or nonlinearity of MIPs, they are designated as mixed-integer linear programs (MILPs) and mixed-integer nonlinear programs (MINLPs).

The principles of optimization theory and algorithms are covered by various books (e.g., Grossmann, 1996; Floudas, 1995; Edgar and Himmelblau, 1988; Reklaitis et al., 1983; Beveridge and Schechter, 1970). Furthermore, several software packages are now commercially available (e.g., LINGO which accompanies this book). It is worth pointing out that most optimization software can efficiently obtain the global solution of LPs and MILPs. On the other hand, no commercial package is guaranteed to identify the global solution of non convex NLPs and MINLPs. Recently, significant research has been undertaken towards developing effective techniques for the global solution of non convex NLPs and MINLPs (e.g., Vaidyanathan and El-Halwagi, 1994, 1996; Sahinidis and Grossmann, 1991; Visweswaran and Floudas, 1990). Within the next few years, these endeavors may indeed lead to practical procedures for globally solving general classes of NLPs and MINLPs.

The optimization component of process integration drives the iterations between synthesis and analysis toward an optimal closure. In many cases, optimization is also used within the synthesis activities. For instance, in the targeting approach for synthesis, the various objectives are reconciled using optimization. In the structure-based synthesis approach, optimization is typically the main framework for formulating and solving the synthesis task.

1.3 Can Flowsheets Provide Global Insights?

Chemical processes involve a strong interaction between mass and energy. Typically, the overall objective of a plant is to convert and process mass. Energy is used to drive reactions, effect separations and drive pumps and compressors. An overview of the main inputs and outputs of a process is shown in Fig. 1.1. The

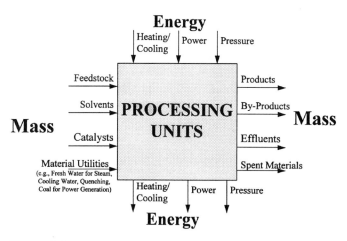

Figure 1.1 The mass-energy matrix of the main inputs and outputs of a process (Garrison et al., 1996).

processing units, the streams and their interconnections are represented by the process flowsheet. Therefore, a process flowsheet illustrates how the raw materials are prepared (e.g., by separation and size reduction), then converted through reaction into desired products, by-products, and wastes that are finally separated and purified. In this regard, a process flowsheet is a valuable tool that allows the engineer to follow the production scheme, conduct the basic material and energy balances and simulate the process performance. But, is that all we need to understand the process? In any plant, there are global insights that characterize the "big picture" for mass and energy flows and hence determine the optimal policies for allocating species (pollutants, products, etc.) and energy throughout the plant. *These global insights are not transparent from a conventional process flowsheet.* To demonstrate this point, let us examine the following motivating example.

Consider a plant which produces ethyl chloride (C_2H_5Cl) by catalytically reacting ethanol and hydrochloric acid (El-Halwagi et al., 1996). Figure 1.2 is a simplified flowsheet of the process. First, ethanol is manufactured by the catalytic hydration of ethylene. Ethanol is separated using distillation followed by membrane separation (pervaporation). Ethanol is reacted with hydrochloric acid to form ethyl chloride. A by-product of the reaction is chloroethanol, CE (C_2H_5OCl), which is a toxic pollutant. The off-gas from the reactor is scrubbed with water in two absorption columns to recover the majority of unreacted ethanol, hydrogen chloride and CE and to purify the product. The aqueous streams leaving the scrubbers are mixed and recycled to the reactor. The aqueous effluent from the ethyl chloride reactor is mixed with the wastewater from the ethanol distillation unit. The terminal wastewater stream is fed to a biotreatment facility for detoxification

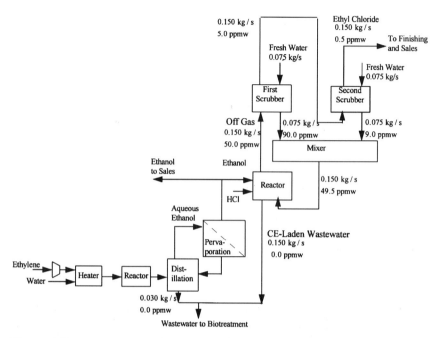

Figure 1.2 A simplified process flowsheet for the production of ethyl chloride with all compositions are in parts per million of CE on a weight basis (El-Halwagi et al., 1996, reproduced with permission of the American Institute of Chemical Engineers. Copyright © 1996 AIChE. All rights reserved).

prior to discharge. Because of the toxicity of CE, it is desired to reduce its amount in terminal wastewater to one-sixth of its current discharge. What would be the minimum-cost solution for this waste-reduction task?

Several strategies are to be considered for reducing the load of discharged CE including source reduction, in-plant separation, and recycle. Six separation processes are candidates for removing CE from aqueous and gaseous streams. For aqueous streams, one may consider adsorption on a polymeric resin, adsorption using activated carbon, and extraction using oil. For removal from gaseous streams, zeolite adsorption, air stripping and steam stripping are potential separations. The thermodynamic data, cost information, and process modeling equations are available (and will be discussed in detail in Chapter Seven).

In order to identify the optimum solution, one should be able to answer the following challenging questions:

· Which phase(s) (gaseous, liquid) should be intercepted with a separation system to remove CE?
· Which process streams should be intercepted?
· To what extent should CE be removed from each process stream to render an overall reduction of 85% in terminal CE loading?

· Which separation operations should be used for interception (e.g., adsorption, extraction, stripping)?
· Which separating agents should be selected for interception (e.g., resin, activated carbon, oil, zeolite, air, steam)?
· What is the optimal flowrate of each separating agent?
· How should these separating agents be matched with the CE-laden streams (i.e., stream pairings)?
· Which units should be manipulated for source reduction? By what means?
· Should any streams be segregated? Which ones?
· Which streams should be recycled/reused? To what units?

To answer the above-mentioned questions, one can envision so many alternatives they cannot be enumerated. Typically, an engineer charged with the responsibility of answering these questions examines few process options based on experience and corporate preference. Consequently, the designer develops a simulation model, performs an economic analysis and selects the least expensive alternative from the limited number of examined options. This solution is inappropriately designated as the "optimum." Normally it is not! Indeed, the true optimum may be an order of magnitude less expensive.

It is beneficial to consider the optimal solution for this case study shown by Fig. 1.3, with the process changes marked in thick lines. The solution features

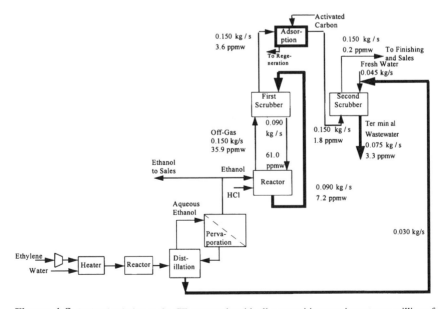

Figure 1.3 Optimal solution to the CE case study with all compositions are in parts per million of CE on a weight basis (El-Halwagi et al., 1996, reproduced with permission of the American Institute of Chemical Engineers. Copyright © 1996 AIChE. All rights reserved).

segregation of aqueous streams, the use of adsorption to remove CE from a *gaseous* stream (although the objective of the problem is to reduce CE loading in the *aqueous* effluent) and the reuse of liquid streams in the scrubbers. This is not an intuitively obvious solution. Nonetheless, it can be generated systematically.

The foregoing discussion illustrates that flowsheets (notwithstanding their usefulness) do not readily provide the global insights of the process. The use of repeated analyses to screen a few arbitrarily generated alternatives can be quite misleading. Instead, what is needed is a systematic methodology that can quickly and smoothly guide engineers through the complexities of the flowsheet, allowing them to identify the big picture of mass and energy flows, determine best performance targets of the process, and extract the optimal solution without having to enumerate and analyze the numerous alternatives. Does such a methodology exist? The answer is yes: via *mass integration and energy integration.*

1.4 Branches of Process Integration: Mass Integration and Energy Integration

As has been discussed earlier, a fundamental understanding of the global flow of mass and energy is instrumental in developing optimal design and operating strategies to meet process objectives including cost effectiveness, yield enhancement, energy efficiency, and pollution prevention. Over the past two decades, significant contributions have been made in understanding the global flow of mass and energy within a process. Two key branches of process integration have been developed: *mass integration and energy integration.* Energy integration is a systematic methodology that provides a fundamental understanding of energy utilization within the process and employs this understanding in identifying energy targets and optimizing heat-recovery and energy-utility systems. Numerous articles on energy integration have been published (for example see reviews by Shenoy, 1995; Linnhoff et al., 1994; Linnhoff, 1993; Gundersen and Naess, 1988). Of particular importance are the thermal-pinch techniques that can be used to identify minimum heating and cooling utility requirements for a process. On the other hand, *mass integration* is a systematic methodology that provides a fundamental understanding of the global flow of mass within the process and employs this holistic understanding in identifying performance targets and optimizing the generation and routing of species throughout the process. Mass-allocation objectives such as pollution prevention are at the heart of mass integration. Mass integration is more general and more involved than energy integration. Because of the overriding mass objectives of most processes, mass integration can potentially provide much stronger impact on the process than energy integration. Both integration branches are compatible. Mass integration coupled with energy integration provides a systematic framework for understanding the big picture of the process, identifying performance targets, and developing solutions for improving process

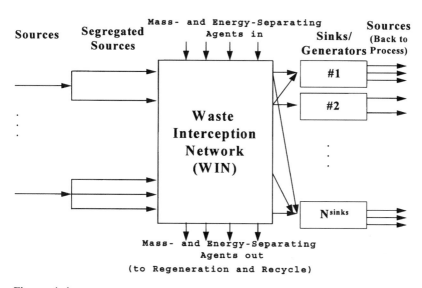

Figure 1.4 Schematic representation of mass-integration strategies for pollution prevention; segregation, mixing, interception, recycle and sink/generator manipulation (El-Halwagi and Spriggs, 1996).

efficiency including pollution prevention. The core of this book is dedicated to mass-integration techniques.

Mass integration is based on fundamental principles of chemical engineering combined with system analysis using graphical and optimization-based tools. The first step in conducting mass integration is the development of a global mass allocation representation of the whole process from a *species viewpoint* (El-Halwagi et al., 1996; El-Halwagi and Spriggs, 1996; Garrison et al., 1995, 1996; Hamad et al., 1995, 1996) as shown in Fig. 1.4. For each targeted species (e.g., each pollutant), there are sources (streams that carry the species) and process sinks (units that can accept the species). Process sinks include reactors, heaters/coolers, biotreatment facilities, and discharge media. Streams leaving the sinks become, in turn, sources. Therefore, sinks are also generators of the targeted species. Each sink/generator may be manipulated via design and/or operating changes to affect the flowrate and composition of what each sink/generator accepts and discharges.

In general, sources must be prepared for the sinks through segregation and separation via a waste-interception network (WIN) (Hamad et al., 1996; El-Halwagi et al., 1995, 1996; Garrison et al., 1995). Effective pollution prevention can be achieved by a combination of stream segregation, mixing, interception, recycle from sources to sinks (with or without interception) and sink/generator manipulation. Therefore, issues such as source reduction and recycle/reuse can be simultaneously addressed (Hamad et al., 1995). The following sections summarize these concepts.

Segregation simply refers to avoiding the mixing of streams. In many cases, segregating waste streams at the source renders several streams environmentally acceptable and hence reduces the pollution-prevention cost. Furthermore, segregating streams with different compositions avoids unnecessary dilution of streams. This reduces the cost of removing the pollutant from the more concentrated streams. It may also provide composition levels that allow the streams to be recycled directly to process units.

Recycle refers to the utilization of a pollutant-laden stream (a source) in a process unit (a sink). Each sink has a number of constraints on the characteristics (e.g., flowrate and composition) of feed that it can process. If a source satisfies these constraints it may be directly recycled to or reused in the sink. However, if the source violates these constraints segregation, mixing, and/or interception may be used to prepare the stream for recycle.

Interception denotes the utilization of separation unit operations to adjust the composition of the pollutant-laden streams to make them acceptable for sinks. These separations may be induced by the use of mass-separating agents (MSAs) and/or energy separating agents (ESAs). A systematic technique is needed to screen the multitude of separating agents and separation technologies to find the optimal separation system. The synthesis of MSA-induced physical-separation systems is referred to as the synthesis of mass-exchange networks (MENs) (El-Halwagi and Manousiouthakis, 1989). Interception networks using reactive MSAs are termed reactive mass exchange networks (REAMEN) (Srinivas and El-Halwagi, 1994; El-Halwagi and Srinivas, 1992). Network synthesis techniques have also been devised for other separation systems that can be used in intercepting pollutants. These systems include pressure-driven membrane separations (e.g., El-Halwagi, 1992; Evangelista, 1986), heat-induced separation networks (HISENs) (e.g., Dunn et al., 1995; Dye et al., 1995; Richburg and El-Halwagi, 1995; El-Halwagi et al., 1995) and distillation sequences (e.g., Malone and Doherty, 1995; Wahnschafft et al., 1991).

Sink/generator manipulation involves design or operating changes that alter the flowrate or composition of pollutant-laden streams entering or leaving the process units. These measures include temperature/pressure changes, unit replacement, catalyst alteration, feedstock substitution, reaction-path changes (e.g., Crabtree and El-Halwagi, 1995), reaction system modification (e.g., Gopalakrishnan et al., 1996; Lakshmanan and Biegler, 1995), and solvent substitution (e.g., Jobak, 1995; Constantinou et al., 1995).

1.5 Structure of the Book

The theory and application of process integration for pollution prevention will be the focus of the rest of the book. Special emphasis is given to mass integration techniques. As has been mentioned in the previous section, pollution prevention can be achieved by a combination of stream segregation, interception, recycle from sources to sinks (with or without interception) and sink/generator manipulation. Segregation and direct recycling opportunities can be readily identified, but interception with recycle and sink/generator manipulation are more challenging issues. Hence, the bulk of the book is dedicated to the systematic identification optimal strategies for interception, intercepted recycle, and sink/generator manipulation. Chapter Two presents an overview of the design of individual MSA-induced separations; referred to as mass-exchange units. Chapters Three, Five, and Six provide graphical, algebraic and optimization techniques for the synthesis of physical MENs. Chapters Four and Seven illustrate how MEN synthesis can be incorporated within a more comprehensive mass-integration analysis. Reactive interceptions are discussed in Chapter Eight. The interaction of heat integration with mass integration is presented in Chapters Nine and Ten. Chapter Eleven focuses on membrane-based interception. Finally, Chapter Twelve briefly discusses the role of chemistry in preventing pollution. All these concepts will be illustrated by a wide variety of case studies.

References

Allen, D. T. and Rosselot, K. S. (1994). Pollution prevention at the macro scale: Flows of wastes, industrial ecology and life cycle analyses. *Waste Manage*, **14**(3–4), 317–328.

Bagajewicz, M. J. and Manousiouthakis, V. (1992). Mass-heat exchange network representation of distillation networks. *AIChE J.*, **38**(11), 1769–1800, .

Beveridge, G. S. G. and Schechter, R. (1970). Optimization: Theory and Practice. McGraw Hill, New York.

Constantinou, L., Jacksland, C., Bagherpour, K., Gani R., and Bogle, L. (1995). Application of group contribution approach to tackle environmentally-related problems. *AIChE Symp. Ser.*, **90**(303):105–116.

Crabtree, E. W. and El-Halwagi, M. M. (1995). Synthesis of environmentally-acceptable reactions. *AIChE Symp. Ser.*, **90**(303), 117–127.

Douglas, J. M. (1992). Process synthesis for waste minimization. *Ind. Eng. Chem. Res.*, **31**, 238–243.

Dunn, R. F., Zhu, M., Srinivas, B. K., and El-Halwagi, M. M. (1995). Optimal design of energy-induced separation networks for VOC recovery. *AIChE Symp. Ser.*, **90**(303), 74–85.

Dye, S. R., Berry, D. A., and Ng, K. M. (1995). Synthesis of crytallization-based separation schemes. *AIChE Symp. Ser.*, **91**(304), 238–241.

Edgar, T. F. and Himmelblau, D. M. (1988). Optimization of chemical processes. McGraw Hill, New York .

El-Halwagi, A. M. and El-Halwagi, M. M. (1992). Waste minimization via computer aided chemical process synthesis—A new design philosophy, *TESCE J.*, **18**(2), 155–187.

El-Halwagi, M. M. (1992). Synthesis of reverse osmosis networks for waste reduction. *AIChE J.*, **38**(8), 1185–1198.

El-Halwagi, M. M. and Manousiouthakis, V. (1989). Synthesis of mass-exchange networks. *AIChE J.*, **35**(8), 1233–1244.

El-Halwagi, M. M. and Petrides, D. M., Eds. (1995). Pollution prevention via process and product modifications. *AIChE Symp. Ser.*, **90**(303), AIChE, New York.

El-Halwagi, M. M. and Spriggs, H. D. (1996). An integrated approach to cost and energy efficient pollution prevention. Proceedings of Fifth World Congr. of Chem. Eng., Vol. III, pp. 344–349, San Diego.

El-Halwagi, M. M. and Srinivas, B. K. (1992). Synthesis of reactive mass exchange networks. *Chem. Eng. Sci.*, **47**(8), 2113–2119.

El-Halwagi, M. M., Srinivas, B. K., and Dunn, R. F. (1995). Synthesis of optimal heat-induced separation networks. *Chem. Eng. Sci.*, **50**(1), 81–97.

El-Halwagi, M. M., Hamad, A. A., and Garrison, G. W. (1996). Synthesis of waste interception and allocation networks. *AIChE J.*, **42**(11), 3087–3101.

Evangelista, F. (1986). Improved graphical analytical method for the design of reverse osmosis desalination plants. *Ind. Eng. Chem. Process Des. Dev.*, **25**(2), 366–375.

Floudas, C. A. (1995). Nonlinear and mixed integer optimization: Fundamentals and applications. Oxford Univ. Press, New York.

Floudas, C. A., Ciric, A. R., and Grossmann, I. E. (1986). Automatic synthesis of optimum heat exchange network configurations. *AIChE J.*, **32**(2), 276–290.

Freeman, H., Ed. (1995). Industrial pollution prevention handbook. McGraw Hill, New York.

Friedler, F., Varga, J. B., and Fan, L. T. (1995). Algorithmic Approach to the integration of total flowsheet synthesis and waste minimization. *AIChE Symp. Ser.*, **90**(303), 86–97. AIChE, NY.

Gopalakrishnan, M., Ramdoss, P., and El-Halwagi, M. (1996). Integrated design of reaction and separation systems for waste minimization. *AIChE Annu. Meet.*, Chicago.

Garrison, G. W., Spriggs, H. D., and El-Halwagi, M. M. (1996). A global approach to integrating environmental, energy, economic and technological objectives. Proceedings of Fifth World Congr. of Chem. Eng., Vol. I, pp. 675–680, San Diego.

Garrison, G. W., Hamad, A. A., and El-Halwagi, M. M. (1995). Synthesis of waste interception networks. *AIChE Annu. Meet.*, Miami.

Grossmann, I. E., Ed. (1996). Global Optimization in Engineering Desig, Kluwer Academic Pub., Dordrecht, The Netherlands.

Hamad, A. A., Garrison, G. W., Crabtree, E. W., and El-Halwagi, M. M. (1996). Optimal design of hybrid separation systems for waste reduction. Proceedings of Fifth World Congr. of Chem. Eng., Vol. III, pp. 453–458, San Diego.

Hamad, A. A., Varma, V., El-Halwagi, M. M., and Krishnagopalan, G. (1995). Systematic integration of source reduction and recycle reuse for the cost-effective compliance with the cluster rules. *AIChE Annu. Meet.*, Miami.

Jobak, K. G. (1995). Solvent substitution for pollution prevention. *AIChE Symp. Ser.*, **90**(303), 98–104, .

Lakshmanan, A. and Biegler, L. T. (1995). Reactor network targeting for waste minimization. *AIChE Symp. Ser.*, **90**(303), 128–138.

Gundersen, T. and Naess, L. (1988). The synthesis of cost optimal heat exchanger networks, an industrial review of the state of the art. *Comput. Chem. Eng.*, **12**(6), 503–530.

Linnhoff, B., Townsend, D. W., Boland, D., Hewitt, G. F., Thomas, B. E. A., Guy, A. R., and Marsland, R. H. (1994). A User Guide on Process Integration for the Efficient Use of Energy. Revised 1st Ed., Institution of Chemical Engineers, Rugby, UK.

Linnhoff, B. (1993). Pinch analysis—A state of the art overview. *Trans. Inst. Chem. Eng. Chem. Eng. Res. Des.*, **71**, Part A5, 503–522.

Malone, M. F. and Doherty, M. F. (1995). Separation system synthesis for nonideal liquid mixtures. *AIChE Symp. Ser.*, **91**(304), 9–18.

Nishida, N., Stephanopoulos, G., and Westerberg, A. (1981). A review of process synthesis. *AIChE J.*, **27**(3), 321–351.

Noyes, R., Ed. (1993). Pollution Prevention Technology Handbook. Noyes Publications, Park Ridge, NJ.

Reklaitis, G. V., Ravindran A., and Ragsdell, K. M. (1983). Engineering Optimization. Wiley, New York.

Richburg, A. and El-Halwagi, M. M. (1995). A graphical approach to the optimal design of heat-induced separation networks for VOC recovery. *AIChE Symp. Ser.*, **91**(304), 256–259.

Sahinidis, N. V. and Grossmann, I. E. (1991). Convergence properties of generalized Benders decomposition. *Comput. Chem. Eng.*, **15**(7), 481–491.

Shenoy, U. V. (1995). "Heat Exchange Network Synthesis: Process Optimization by Energy and Resource Analysis." Gulf Pub. Co., Houston, TX.

Srinivas, B. K. and El-Halwagi, M. M. (1994). Synthesis of reactive mass-exchange networks with general nonlinear equilibrium functions. *AIChE J.*, **40**(3), 463–472.

Stephanopoulos, G. and Townsend, D. (1986). Synthesis in process development. *Chem. Eng. Res. Des.*, **64**(3), 160–174.

Theodore, L., Dupont, R. R., and Reynolds, J., Eds. (1994). "Pollution Prevention: Problems and Solutions." Gordon & Breach, Amsterdam.

Vaidyanathan, R. and El-Halwagi, M. M. (1996). Global optimization of nonconvex MINLP's by interval analysis. "In Global Optimization in Engineering Design," (I. E. Grossmann, ed.), pp. 175–194. Kluwer Academic Publishers, Dordrecht, The Netherlands.

Vaidyanathan, R. and El-Halwagi, M. M. (1994). Global optimization of nonconvex nonlinear programs via interval analysis. *Comput. Chem. Eng.*, **18**(10), 889–897.

Visweswaran, V. and Floudas, C. A. (1990). A global optimization procedure for certain classes of nonconvex NLP's–II. application of theory and test problems. *Comput. Chem. Eng.*, **14**(2), 1419–1434.

Wahnschafft, O. M., Jurian, T. P., and Westerberg, A. W. (1991). SPLIT: A separation process designer. *Comput. Chem. Eng.*, **15**, 565–581.

Westerberg, A. W. (1987). Process synthesis: A morphological view. In Recent Developments in Chemical Process and Plant Design, (Y. A. Liu, H. A. McGee, Jr., and W. R. Epperly, eds), pp. 127–145. Wiley, New York.

CHAPTER TWO

Modeling of Mass-Exchange Units
for Environmental Applications

Mass-exchange operations are indispensable for pollution prevention. Within a mass-integration framework, mass-exchange operations are employed in intercepting sources by selectively transferring certain undesirable species from a number of waste streams (sources) to a number of mass-separating agents (MSAs). The objective of this chapter is to provide an overview of the basic modeling principles of mass-exchange units. For a more comprehensive treatment of the subject, the reader is referred to McCabe et al. (1993), Wankat (1988), Geankoplis (1983), Henley and Seader (1981), King (1980) and Treybal (1980).

2.1 What Is a Mass Exchanger?

A mass exchanger is any direct-contact mass-transfer unit that employs an MSA (or a lean phase) to selectively remove certain components (e.g., pollutants) from a rich phase (e.g., a waste stream). The MSA should be partially or completely immiscible in the rich phase. When the two phases are in intimate contact, the solutes are redistributed between the two phases leading to a depletion of the rich phase and an enrichment of the lean phase. Although various flow configurations may be adopted, emphasis will be given to countercurrent systems because of their industrial importance and efficiency. The realm of mass exchange includes the following operations:

Absorption, in which a liquid solvent is used to remove certain compounds from a gas by virtue of their preferential solubility. Examples of absorption involve desulfurization of flue gases using alkaline solutions or ethanolamines, recovery of volatile-organic using light oils, and removal of ammonia from air using water.

Adsorption, which utilizes the ability of a solid adsorbent to adsorb specific components from a gaseous or a liquid solution onto its surface. Examples of adsorption include the use of granular activated carbon for the removal of benzene/toluene/xylene mixtures from underground water, the separation of ketones from aqueous wastes of an oil refinery, and the recovery of organic solvents from the exhaust gases of polymer manufacturing facilities. Other examples include the use of activated alumina to adsorb fluorides and arsenic from metal-finishing emissions.

Extraction, employs a liquid solvent to remove certain compounds from another liquid using the preferential solubility of these solutes in the MSA. For instance, wash oils can be used to remove phenols and polychlorinated biphenyls (PCBs) from the aqueous wastes of synthetic-fuel plants and chlorinated hydrocarbons from organic wastewater.

Ion exchange, in which cation and/or anion resins are used to replace undesirable anionic species in liquid solutions with nonhazardous ions. For example, cation-exchange resins may contain nonhazardous, mobile, positive ions (e.g., sodium, hydrogen) which are attached to immobile acid groups (e.g., sulfonic or carboxylic). Similarly, anion-exchange resins may include nonhazardous, mobile, negative ions (e.g., hydroxyl or chloride) attached to immobile basic ions (e.g., amine). These resins can be used to eliminate various species from wastewater, such as dissolved metals, sulfides, cyanides, amines, phenols, and halides.

Leaching, which is the selective solution of specific constituents of a solid mixture when brought in contact with a liquid solvent. It is particularly useful in separating metals from solid matrices and sludge.

Stripping, which corresponds to the desorption of relatively volatile compounds from liquid or solid streams using a gaseous MSA. Examples include the recovery of volatile organic compounds from aqueous wastes using air, the removal of ammonia from the wastewater of fertilizer plants using steam, and the regeneration of spent activated carbon using steam or nitrogen.

2.2 Equilibrium

Consider a lean phase, j, which is in intimate contact with a rich phase, i, in a closed vessel in order to transfer a certain solute. The solute diffuses from the rich phase to the lean phase. Meanwhile, a fraction of the diffused solute back-transfers to the rich phase. Initially, the rate of rich-to-lean solute transfer surpasses that of lean to rich leading to a net transfer of the solute from the rich phase to the lean phase. However, as the concentration of the solute in the rich phase increases, the back-transfer rate also increases. Eventually, the rate of rich-to-lean solute transfer becomes equal to that of lean to rich, resulting in a dynamic equilibrium with zero net interphase transfer. Physically, this situation corresponds to the state at which both phases have the same value of chemical potential for the solute. In the case of ideal systems, the transfer of one component is indifferent to the transfer of other species. Hence, the composition of the solute in the rich phase, y_i, can be related to its composition in the lean phase, x_j, via an equilibrium distribution function, f_j^*, which is a function of the system characteristics including temperature and pressure. Hence, for a given rich-stream composition, y_i, the maximum attainable composition of the solute in the lean phase, x_j^*, is given by

$$y_i = f_j^*\left(x_j^*\right). \tag{2.1}$$

Many environmental applications involve dilute systems whose equilibrium functions can be linearized over the operating range to yield

$$y_i = m_j x_j^* + b_j. \tag{2.2}$$

Special cases of Eq. (2.2) include Raoult's law for absorption:

$$y_i = \frac{P_{solute}^o(T)}{P_{total}} x_j^*. \tag{2.3}$$

where y_i and x_j^* are the mole fractions of the solute in the gas and the liquid phases, respectively, $p_{solute}^o(T)$ is the vapor pressure of the solute at a temperature T and P_{total} is the total pressure of the gas.

Another example of Eq. (2.2) is Henry's law for stripping:

$$y_i = H_j x_j^*, \tag{2.4}$$

where y_i and x_j^* are the mole fractions of the solute in the liquid waste and the stripping gas, respectively,[1] and H_j is Henry's coefficient, which may be theoretically approximated by the following expression:

$$H_j = \frac{P_{total} \cdot y_i^{solubility}}{P_{solute}^o(T)}, \tag{2.5}$$

in which $p_{solute}^o(T)$ is the vapor pressure of the solute at a temperature T, P_{total} is the total pressure of the stripping gas, and $y_i^{solubility}$ is the liquid-phase solubility of

---

[1] Throughout this book, several mass-exchange operations will be considered simultaneously. It is therefore necessary to use a unified terminology such that y is always the composition in the rich phase and x is the composition in the lean phase. The reader is cautioned here that this terminology may be different from other literature, in which y is used for gas-phase composition and x is used for liquid-phase composition.

the pollutant at temperature T (expressed as mole fraction of the pollutant in the liquid waste).

An additional example of Eq. (2.2) is the distribution function commonly used in solvent extraction:

$$y_i = K_j x_j^*, \tag{2.6}$$

where y_i and x_j^* are the compositions of the pollutant in the liquid waste and the solvent, respectively, and K_j is the distribution coefficient.

Accurate experimental results provide the most reliable source for equilibrium data. If not available, empirical correlations for predicting equilibrium data may be invoked. These correlations are particularly useful at the conceptual-design stage. Several literature sources provide compilations of equilibrium data and correlations: for example Lo et al. (1983) as well as Francis (1963) for solvent extraction, Reid et al. (1987) for vapor-liquid and liquid-liquid systems, Hwang et al. (1992) for steam stripping, Mackay and Shiu (1981), Fleming (1989), Clark et al. (1990) and Yaws (1992) for air stripping, U.S. Environmental Protection Agency (1980), Cheremisinoff and Ellerbusch (1980), Perrich (1981), Yang (1987), Valenzuela and Myers (1989), Stenzel and Merz (1989), Stenzel (1993), and Yaws et al. (1995) for adsorption, Kohl and Riesenfeld (1985) for gas adsorption and absorption and Astarita et al. (1983) for reactive absorption.

2.3 Interphase Mass Transfer

Whenever the rich and the lean phases are not in equilibrium, an interphase concentration gradient and a mass-transfer driving force develop leading to a net transfer of the solute from the rich phase to the lean phase. A common method of describing the rates of interphase mass transfer involves the use of overall mass-transfer coefficients which are based on the difference between the bulk concentration of the solute in one phase and its equilibrium concentration in the other phase. Suppose that the bulk concentrations of a pollutant in the rich and the lean phases are y_i and x_j, respectively. For the case of linear equilibrium, the pollutant concnetration in the lean phase which is in equilibrium with y_i is given by

$$x_j^* = (y_i - b_j)/m_j \tag{2.7}$$

and the pollutant concentration in the rich phase which is in equilibrium with x_j can be represented by

$$y_i^* = m_j x_j + b_j \tag{2.8}$$

Let us define two overall mass transfer coefficients; one for the rich phase, K_y, and one for the lean phase, K_x. Hence, the rate of interphase mass transfer for the pollutant, $N_{pollutant}$, can be defined as,

$$N_{pollutant} = K_y \left(y_i - y_i^* \right) \tag{2.9a}$$

$$= K_x \left(x_j^* - x_j \right) \tag{2.9b}$$

Correlations for estimating overall mass-transfer coefficients can be found in McCabe et al. (1993), Perry and Green (1984), Geankoplis (1983), Henley and Seader (1981), King (1980) and Treybal (1980).

2.4 Types and Sizes of Mass Exchangers

The main objective of a mass exchanger is to provide appropriate contact surface for the rich and the lean phases. Such contact can be accomplished by using various types of mass-exchange units and internals. In particular, there are two primary categories of mass-exchange devices: *multistage and differential contactors*. In a *multistage* mass exchanger, each stage provides intimate contact between the rich and the lean phases followed by phase separation. Because of the thorough mixing, the pollutants are redistributed between the two phases. With sufficient mixing time, the two phases leaving the stage are essentially in equilibrium; hence the name equilibrium stage. Examples of a multiple-stage mass exchanger include tray columns (Fig. 2.1) and mixers settlers (Fig. 2.2).

In order to determine the size of a multiple-stage mass exchanger, let us consider the isothermal mass exchanger shown in Fig. 2.3. The rich (waste) stream, i,

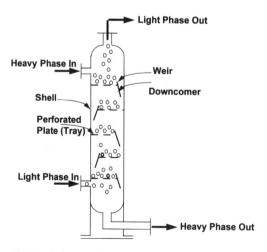

Figure 2.1 A multistage tray column.

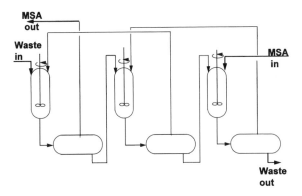

Figure 2.2 A three-stage mixer-settler system.

has a flow rate G_i and its content of the pollutant must be reduced from an inlet composition, y_i^{in}, to an outlet composition, y_i^{out}. An MSA (lean stream), j, (whose flowrate is L_j, inlet composition is x_j^{in} and outlet composition is x_j^{out}) flows countercurrently to selectively remove the pollutant.[2] Figure 2.4 is a schematic representation of the multiple stages of this mass exchanger. If the n th block is an equilibrium stage, then the compositions $y_{i,n}$ and $x_{j,n}$ are in equilibrium. On the other hand, the two compositions on the same end of the stage (e.g., $y_{i,n-1}$ and $x_{j,n}$) are said to be operating with each other.

One way of calculating the number of equilibrium stages (or number of theoretical plates, NTP) for a mass exchanger is the graphical McCabe-Thiele method. To illustrate this procedure, let us assume that over the operating range of compositions, the equilibrium relation governing the transfer of the pollutant from the

---

[2]Once again, because of the unified approach of this text to all mass-exchange operations it is important to emphasize that the symbols G and L will be used to designate the flowrates of the rich stream and the MSA, respectively and not necessarily flowrates of gas and liquid.

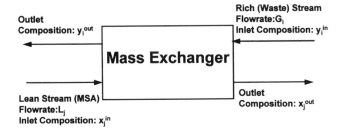

Figure 2.3 A mass exchanger.

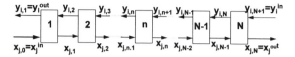

Figure 2.4 A schematic diagram of a multistage mass exchanger.

waste stream to the MSA can be represented by the linear expression described by Eq. (2.2). A material balance on the pollutant that is transferred from the waste stream to the MSA may be expressed as

$$G_i\left(y_i^{in} - y_i^{out}\right) = L_j\left(x_j^{out} - x_j^{in}\right). \tag{2.10}$$

On a y-x (McCabe-Thiele) diagram, this equation represents the operating line which extends between the points (y_i^{in}, x_j^{out}) and (y_i^{out}, x_j^{in}) and has a slope of L_j/G_i, as shown in Fig. 2.5. Furthermore, each theoretical stage can be represented by a step between the operating line and the equilibrium line. Hence, NTP can be determined by "stepping off" stages between the two ends of the exchanger, as illustrated by Fig. 2.5.

Alternatively, for the case of isothermal, dilute mass exchange with linear equilibrium, NTP can be determined through the Kremser (1930) equation:

$$NTP = \frac{\ln\left[\left(1 - \frac{m_j G_i}{L_j}\right)\left(\frac{y_i^{in} - m_j x_j^{in} - b_j}{y_i^{out} - m_j x_j^{in} - b_j}\right) + \frac{m_j G_i}{L_j}\right]}{\ln\left(\frac{L_j}{m_j G_i}\right)} \tag{2.11}$$

Other forms of the Kremser equation include

$$NTP = \frac{\ln\left[\left(1 - \frac{L_j}{m_j G_i}\right)\left(\frac{x_i^{in} - x_j^{out,*}}{x_j^{out} - x_j^{out,*}}\right) + \frac{L_i}{m_j G_i}\right]}{\ln\left(\frac{m_j G_i}{L_j}\right)}, \tag{2.12}$$

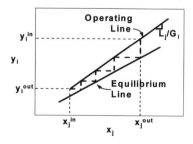

Figure 2.5 The McCabe Thiele diagram.

where

$$x_j^{out,*} = \frac{y_i^{in} - b_j}{m_j}. \tag{2.13}$$

Also,

$$\frac{y_i^{in} - m_j x_j^{out} - b_j}{y_i^{out} - m_j x_j^{in} - b_j} = \left(\frac{L_j}{m_j G_i} \right)^{NTP}. \tag{2.14}$$

If contact time is not enough for each stage to reach equilibrium, one may calculate the number of actual plates "NAP" by incorporating contacting efficiency. Two principal types of efficiency may be employed: overall and stage. The overall exchanger efficiency, η_o, can be used to relate NAP and NTP as follows

$$NAP = NTP/\eta_o. \tag{2.15}$$

The stage efficiency may be defined based on the rich phase or the lean phase. For instance, when the stage efficiency is defined for the rich phase, η_y, Eq. (2.11) becomes

$$NTP = \frac{\ln \left[\left(1 - \frac{m_j G_i}{L_j} \right) \left(\frac{y_i^{in} - m_j x_j^{in} - b_j}{y_i^{out} - m_j x_j^{in} - b_j} \right) + \frac{m_j G_i}{L_j} \right]}{-\ln \left\{ 1 + \eta_y \left[\left(\frac{m_j G_i}{L_j} \right) - 1 \right] \right\}} \tag{2.16}$$

The second type of mass-exchange units is the ***differential (or continuous)*** contactor. In this category, the two phases flow through the exchanger in continuous contact throughout without intermediate phase separation and recontacting. Examples of differential contactors include packed columns (Fig. 2.6), spray towers (Fig. 2.7), and mechanically agitated units (Fig. 2.8).

The height of a differential contactor, H, may be estimated using

$$H = HTU_y NTU_y \tag{2.17a}$$

$$= HTU_x NTU_x, \tag{2.17b}$$

where HTU_y and HTU_x are the overall height of transfer units based on the rich and the lean phases, respectively, while NTU_y and NTU_x are the overall number of transfer units based on the rich and the lean phases, respectively. The overall height of a transfer unit may be provided by the packing (or unit) manufacturer or estimated using empirical correlations (typically by dividing superficial velocity of one phase by its overall mass transfer coefficient). On the other hand, the number of transfer units can be theoretically estimated for the case of isothermal, dilute

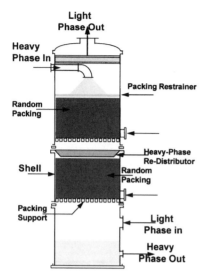

Figure 2.6 Schematic diagram of a countercurrent packed column.

mass exchangers with linear equilibrium as follows:

$$NTU_y = \frac{y_i^{in} - y_i^{out}}{\left(y_i - y_i^*\right)_{\log mean}}, \qquad (2.18a)$$

where

$$\left(y_i - y_i^*\right)_{\log mean} = \frac{\left(y_i^{in} - m_j x_j^{out} - b_j\right) - \left(y_i^{out} - m_j x_j^{in} - b_j\right)}{\ln\left(\frac{\left(y_i^{in} - m_j x_j^{out} - b_j\right)}{\left(y_i^{out} - m_j x_j^{in} - b_j\right)}\right)} \qquad (2.18b)$$

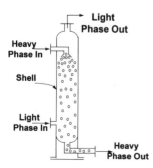

Figure 2.7 A spray column.

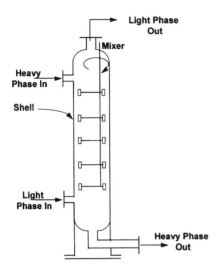

Light Phase
Out

Mixer

Heavy
Phase In

Shell

Light
Phase In

Heavy Phase
Out

Figure 2.8 A mechanically-agitated mass exchanger.

and

$$NTU_x = \frac{x_j^{in} - x_j^{out}}{\left(x_j - x_j^*\right)_{\log mean}}, \qquad (2.19a)$$

where

$$\left(x_j - x_j^*\right)_{\log mean} = \frac{\left[x_j^{out} - \left(\dfrac{y_i^{in} - b_j}{m_j}\right)\right] - \left[x_j^{in} - \left(\dfrac{y_i^{out} - b_j}{m_j}\right)\right]}{\ln\left\{\dfrac{\left[x_j^{out} - \left(\dfrac{y_i^{in} - b_j}{m_j}\right)\right]}{\left[x_j^{in} - \left(\dfrac{y_i^{out} - b_j}{m_j}\right)\right]}\right\}}. \qquad (2.19b)$$

If the terminal compositions or L_j/G_i are unknown, it is convenient to use the following form:

$$NTU_y = \frac{\ln\left[\left(1 - \dfrac{m_j G_i}{L_j}\right)\left(\dfrac{y_i^{in} - m_j x_j^{in} - b_j}{y_i^{out} - m_j x_j^{in} - b_j}\right) + \dfrac{m_j G_i}{L_j}\right]}{1 - \left(\dfrac{m_j G_i}{L_j}\right)}. \qquad (2.20)$$

The column diameter is normally determined by selecting a superficial velocity for one (or both) of the phases. This velocity is intended to ensure proper mixing while avoiding hydrodynamic problems such as flooding, weeping, or entrainment. Once a superficial velocity is determined, the cross-sectional area of the column is obtained by dividing the volumetric flowrate by the velocity.

2.5 Minimizing Cost of Mass-Exchange Systems

Appendix III provides an overview of process economics. Two principal categories of expenditure are particularly important; fixed and operating costs. The fixed cost ($) can be distributed over the service life of the equipment as an annualized fixed cost ($/yr). The total annualized cost (TAC) of the system is given by

$$\text{TAC} = \text{annual operating cost} + \text{annualized fixed cost.} \qquad (2.21)$$

In order to minimize TAC, one can iteratively vary mass-transfer driving force to trade off annualized fixed cost versus annual operating cost. In order to demonstrate this concept, let us consider the isothermal mass exchanger shown in Fig. 2.3. The rich (waste) stream, i, has a given flow rate G_i and a known inlet composition of pollutant y_i^{in}. The outlet composition of the pollutant in the waste stream is denoted by y_i^{out}. An MSA (lean stream) j which has a given inlet composition x_j^{in} but whose flowrate L_j is not fixed flows countercurrently to selectively remove the pollutant. Consider that over the operating range of compositions, the equilibrium relation governing the transfer of the pollutant from the waste stream to the MSA can be represented by the linear expression in Eq. (2.2).

A material balance on the pollutant that is transferred from the waste stream to the MSA may be expressed as in Eq. (2.10):

$$G_i\left(y_i^{in} - y_i^{out}\right) = L_j\left(x_j^{out} - x_j^{in}\right)$$

On a y-x (McCabe-Thiele) diagram, this equation represents the operating line that extends between the points (y_i^{in}, x_j^{out}) and (y_i^{out}, x_j^{in}) and has a slope of L_j/G_i. Equation (2.10) has three unknowns, y_i^{out}, L_j and x_j^{out}. Let us first fix the outlet composition of the waste stream, y_i^{out}. As depicted by Fig. 2.9, the maximum theoretically attainable outlet composition of the MSA, $x_j^{out,*}$, is in equilibrium with y_i^{in}. An infinitely large mass exchanger, however, will be needed to undertake

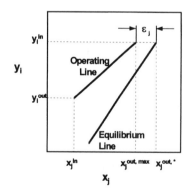

Figure 2.9 Minimum allowable composition difference at the rich end of a mass exchanger.

this mass-transfer duty because of the vanishing mass-exchange driving force at the rich end of the unit. Hence, it is necessary to assign a minimum driving force between the operating and the equilibrium lines at the rich end of the exchanger (Fig. 2.9) such that

$$x_j^{out,\,max} = x_j^{out,*} - \varepsilon_j \tag{2.22}$$

but

$$y_i^{in} = m_j x_j^{out,*} + b_j \tag{2.23}$$

Combining Eqs. (2.22) and (2.23), one obtains

$$x_j^{out,max} = \frac{y_i^{in} - b_j}{m_j} - \varepsilon_j \tag{2.24}$$

where ε_j is referred to as the "minimum allowable composition difference" and $x_j^{out,\,max}$ is the maximum practically feasible outlet composition of the MSA which satisfies the assigned driving force, ε_j.

On the other hand, when x_j^{out} is fixed and y_j^{out} is left as an unknown, the minimum theoretically attainable inlet composition of the waste stream, $y_i^{out,*}$ is in equilibrium with x_j^{in} (Fig. 2.10), i.e.

$$y_i^{out,*} = m x_j^{in} + b_j \tag{2.25}$$

By employing a minimum allowable composition difference of ε_j, at the lean end of the exchanger, one can identify the minimum practically feasible outlet composition of the waste stream to be $y_j^{out,\,min}$ which is given by

$$y_i^{out,\,min} = m_j\left(x_j^{in} + \varepsilon_j\right) + b_j \tag{2.26}$$

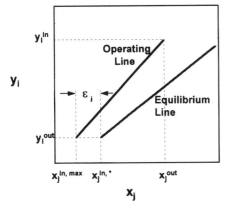

Figure 2.10 Minimum allowable composition difference at the lean end of a mass exchanger.

The selection of the mass-transfer driving forces throughout the exchanger determines the trade-off between the fixed and the operating costs of the system. In many cases (particularly when the equilibrium line is either linear or convex which is typical in environmental applications), the mass-transfer driving forces at both ends of the exchanger can be used to characterize the driving forces throughout the column. Therefore, the parameter ε_j provides a convenient way for trading off fixed versus operating costs of a mass exchanger. For instance, for given values of the inlet and outlet compositions of the rich stream and inlet composition of the lean stream, the value of ε_j, at the rich end of the exchanger can be used to trade off fixed versus operating costs. As ε_j, at the rich end of the exchanger is increased, the slope of the operating line (L_j/G_i) becomes larger and, consequently, the required flowrate of the MSA increases for a given value of the flowrate of the waste stream. Typically, the flowrate of the MSA is the most important element of the operating cost of a mass exchanger. As the flowrate of the MSA increases, additional makeup, regeneration and materials handling (pumping, compression, etc.) are needed leading to an increase in the operating cost of the system. On the other hand, when the value of ε_j at the rich end is increased, the column size (number of stages, height, etc.) decreases which typically results in a reduction in the fixed cost of the mass exchanger. Hence, the designer can iteratively vary ε_j at the rich end until the minimum TAC is identified (see Example 2.2).

Example 2.1: Removal of Toluene from an Aqueous Waste

Air stripping is used to remove 90% of the toluene (molecular weight $= 92$) dissolved in a 10 kg/s (159 gpm) wastewater stream. The inlet composition of toluene in the wastewater is 500 ppm. Air (essentially free of toluene) is compressed to 202.6 kPa (2 atm) and bubbled through a stripper which contains sieve trays. In order to avoid fire hazards, the concentration of toluene in the air leaving the stripper is taken as 50% of the lower flammability limit (LFL) of toluene in air. The toluene-laden air exiting the stripper is fed to a condenser which recovers almost all the toluene. A schematic representation of the process is shown in Fig. 2.11. Calculate the annual operating cost and the fixed capital investment for the system.

The following physical and economic data are available:

Physical data

- The stripping operation takes place isothermally at 298 K and follows Henry's law.
- The vapor pressure of toluene at 298 is 3.8 kPa.
- The solubility of toluene in water at 298 K is 9.8×10^{-3} mol/mol%.
- Lower flammability limit for toluene in air is 12,000 ppm (Crowl and Louvar, 1990).

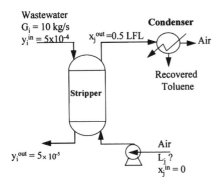

Wastewater
$G_i = 10$ kg/s
$y_i^{in} = 5 \times 10^{-4}$
$x_j^{out} = 0.5$ LFL
Condenser
Air
Recovered
Toluene
Stripper
$y_i^{out} = 5 \times 10^{-5}$
Air
L_j?
$x_j^{in} = 0$

Figure 2.11 Stripping of toluene from wastewater.

Stripper sizing criteria

- The overall efficiency of the column, η_o, is 0.25.
- Sieve plates are used with 0.41-m (16-inch) tray spacing.
- Height of column shell:

$$H = 1.3 \times \text{number of trays} \times \text{tray spacing} \qquad (2.27)$$

(i.e., 30% additional height is used to allow for gas distribution, wastewater feed, gas-liquid disengagement, additional surge volume for wastewater, etc.).
- The maximum allowable superficial velocity of air in the stripper can be estimated using the following expression (based on the data provided by Peters and Timmerhaus (1991) for 0.41-m tray spacing):

$$\text{Maximum allowable superficial velocity of air (m/s)} = 0.068\sqrt{\frac{\rho_{water} - \rho_{air}}{\rho_{air}}}.$$

$$(2.28)$$

Operating cost

- Operating cost is primarily attributed to air compression and toluene condensation
- The operating cost for air compression is basically the electric utility needed for the isentropic compression. Electric energy needed to compress air may be calculated using the following expression:

Compression *Energy* (kJ/kg)

$$= \left(\frac{\gamma}{\gamma - 1}\right)\left(\frac{RT_{in}}{M_{air}\eta_{isentropic}}\right)\left[\left(\frac{P_{out}}{P_{in}}\right)^{\left(\frac{\gamma-1}{\gamma}\right)} - 1\right], \qquad (2.29)$$

where $\gamma = 1.4$ and $\eta_{isentropic} = 0.60$. The electric energy cost is $0.04/kWhr.

- The operating cost for the refrigerant needed for toluene condensation is 6.0×10^{-3}/ kg of toluene-laden air.
- The system is to be operated for 8000 hr/annum.

Equipment cost

- The cost, $, of the stripper (including installation and auxiliaries, but excluding the sieve trays) is given by

$$\text{Cost of column shell} = 1,800 H^{0.85} D^{0.95} \qquad (2.30)$$

where H is the column height (m) and D is the column diameter (m).
- The installed cost of a sieve tray, $, is

$$\text{Cost} = 800 D^{0.86} \qquad (2.31)$$

where D is the column diameter (m).
- The installed cost the air blower, $, is

$$\text{Cost} = 12,000 L_j^{0.6} \qquad (2.32)$$

where L_j is the flowrate of air (kg/s).
- The installed cost of the condenser, $, is

$$\text{Cost} = 62,000 L_j^{0.6} \qquad (2.33)$$

where L_j is the flowrate of air (kg/s).

Solution Let us first determine the values of some of the physical properties needed in solving the problem.

The air density can be determined as follows (see Appendix I):

$$\rho_{air} = \frac{P M_{air}}{RT}. \qquad (2.34)$$

Therefore, density of air at 2 atm and 298 K is

$$\rho_{air} = \frac{2 \times 29}{0.082057 \times 298} \approx 2.4 \, kg/m^3.$$

Since the problem will be worked out in mass units, it is necessary to evaluate compositions and Henry's constant on a mass basis.

The LFL of toluene is 12,000 ppm, which corresponds to a mole fraction of 0.012 (see Appendix I) and can be converted into mass fraction as follows:

$$\text{LFL (in mass fraction units)} = 0.012 \times \frac{92}{29}$$

$$= 0.038 \, \text{kg toluene/kg air.}$$

Since the outlet composition of toluene in air is taken as 50% of the LFL, then

$$x_j^{out} = 0.5 \times 0.038$$

$$= 0.019 \, \text{kg toluene/kg air.}$$

Inlet composition of wastewater is 500 ppm, which corresponds to a mass fraction of $y_i^{in} = 5 \times 10^{-4}$ (see Appendix I). Since 90% of the toluene is removed from the feed, $y_i^{out} = 5 \times 10^{-5}$.

The component material balance on toluene is

$$L_j(0.019 - 0.000) = 10(5 \times 10^{-4} - 5 \times 10^{-5})$$

i.e., $$L_j = 0.237 \, \text{kg air/s.}$$

Using Eq. (2.5), one may estimate Henry's constant in molar units (mole fraction of toluene in water/mole fraction of toluene in air) to be

$$\frac{9.8 \times 10^{-5} \times 202.6}{3.8} = 5.22 \times 10^{-3} \text{ mole fraction of toluene in water/}$$
$$\text{mole fraction of toluene in air}$$

which can be converted into mass units (mass fraction of toluene in water/mass fraction of toluene in air) as follows:

$$m_j = 5.22 \times 10^{-3} \left(\frac{29}{18}\right)$$

$$= 0.0084 \, \frac{\text{mass fraction of toluene in water}}{\text{mass fraction of toluene in air}}.^3$$

---

[3] This value is in good agreement with the experimental constant reported by Mackay and Shiu (1981) to be $0.673 \, \text{kPa} \cdot \text{m}^3/\text{gm mol}$. It is instructive to demonstrate the conversion between different ways of reporting Henry's coefficient. First, the reported value is inverted to be in the units of composition in the rich phase divided by composition in the lean phase, i.e., $1.486 \, \text{gm mol/(kPa} \cdot \text{m}^3)$ which can be converted into units of mole fraction as follows:

$$1.486 \left(\frac{gm \, mol \, toluene}{kPa \, toluene \cdot m^3 water}\right)(202.6 \, kPa \, air)\left(\frac{m^3 water}{10^6 gm \, water}\right)\left(\frac{18 \, gm \, water}{gm \, mol \, water}\right)$$

$$= 5.419 \times 10^{-3} \, \frac{\text{mole fraction of toluene in water}}{\text{mole fraction of toluene in air}}$$

Hence, in terms of mass fraction units

$$5.419 \times 10^{-3} \frac{\text{mole fraction of toluene in water}}{\text{mole fraction of toluene in air}} \times \frac{29 \, \text{kg air}}{\text{kg mol air}} \times \frac{\text{kg mol water}}{18 \, \text{kg water}}$$

$$= 0.0087 \, \frac{\text{mass fraction of toluene in water}}{\text{mass fraction of toluene in air}}.$$

Column sizing We are now in a position to size the column. When Eq. (2.11) is employed, we get

$$NTP = \frac{\ln\left[\left(1 - \dfrac{0.0084 \times 10}{0.237}\right)\left(\dfrac{5 \times 10^{-4} - 0}{5 \times 10^{-5} - 0}\right) + \dfrac{0.0084 \times 10}{0.237}\right]}{\ln\left(\dfrac{0.237}{0.0084 \times 10}\right)}$$

$$= 1.85 \text{ stages}.$$

Therefore,

$$NAP = 1.85/0.25$$
$$= 7.4.$$

i.e., we need eight stages and the column height can now be evaluated using Eq. (2.27)

$$H = 1.3 \times 8 \times 0.41$$
$$= 4.3 \text{ m}.$$

Using Eq. (2.28), we get

$$\text{Maximum allowable superficial velocity of air (m/s)} = 0.068\sqrt{\frac{1000 - 2.4}{2.4}}$$
$$= 1.386 \text{ m/s}.$$

But

Minimum column diameter

$$= \sqrt{\frac{4(\textit{Volumetric Flowrate of Air})}{\pi(\textit{Maximum Allowable Superficial Velocity of Air})}}$$

$$= \sqrt{\frac{4 \times 0.237/2.4}{3.14 \times 1.386}}$$

$$= 0.3 \text{ m}.$$

Now that the size of the column has been determined, it is possible to evaluate the equipment cost.

Estimation of equipment cost Using Eqs. (2.30)–(2.33), we get

$$\text{Installed Equipment Cost} = (1,800)(4.3)^{0.85}(0.3)^{0.95} + (8)(800)(0.3)^{0.86}$$
$$+ (12,000 + 62,000)(0.237)^{0.6}$$
$$= \$35,450.$$

As for operating cost, according to Eq. (2.29),

$$\text{Compression energy} = \left(\frac{1.4}{1.4 - 1}\right)\left(\frac{8.314 \times 300}{29 \times 0.6}\right)\left[\left(\frac{2}{1}\right)^{\left(\frac{1.4-1}{1.4}\right)} - 1\right]$$

$$= 110 \text{ kJ/kg of air.}$$

Since the electric energy cost is $0.04/kWhr, then

$$\text{Compression cost} = 110\frac{\text{kJ}}{\text{kg}} \times \frac{\text{kWhr}}{3600 \text{ kJ}} \times \frac{\$0.04}{\text{kWhr}}$$

$$= 1.2 \times 10^{-3} \text{ \$/kg of air.}$$

But, operating cost is primarily attributed to compression and condensation. Hence,

Annual operating cost

$$= (1.2 \times 10^{-3} + 6.0 \times 10^{-3})\frac{\$}{\text{kg air}} \times 0.237\frac{\text{kg air}}{\text{s}} \times 3600 \times 8000\frac{\text{s}}{\text{yr}}$$

$$= \$49,100/\text{yr.}$$

It is worth pointing out that the value of recovered toluene is $25,900/yr.

Example 2.2: Recovery of Benzene from a Gaseous Emission

Benzene is to be removed from a gaseous emission by contacting it with an absorbent (wash oil, molecular weight 300). The gas flowrate is 0.2 kg-mol/s (about 7,700 ft^3/min) and it contains 0.1 mol/mol% (1000 ppm) of benzene. The molecular weight of the gas is 29, its temperature is 300 K, and it has a pressure of 141 kPa (approximately 1.4 atm). It is desired to reduce the benzene content in the gas to 0.01 mol/mol% using the system shown in Fig. 2.12. Benzene is first absorbed into oil. The oil is then fed to a regeneration system in which oil is heated and passed to a flash column that recovers benzene as a top product. The bottom product is the regenerated oil, which contains 0.08 mol/mol% benzene. The regenerated oil is cooled and pumped back to the absorber.

What is the optimal flowrate of recirculating oil that minimizes the TAC of the system?

The following data may be used

Equilibrium data

· The absorption operation is assumed to be isothermal (at 300 K) and to follow Raoult's law.

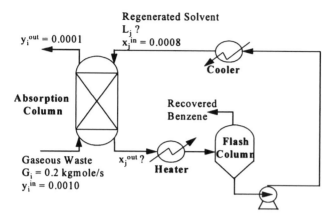

Figure 2.12 Recovery of benzene from a gaseous emission.

- The vapor pressure of benzene, p^o, is given by:

$$\ln p^o = 20.8 - \frac{2789}{T - 52},$$ (2.35)

where p^o is in Pascals and T is in Kelvins.

Absorber sizing criteria

- The overall-gas height of transfer unit for the packing is 0.6 m.
- The superficial velocity of the gas in the absorber is taken as 1.5 m/s to avoid flooding.
- The mass velocity of oil in the absorber should be kept above 2.7 kg/m²· s to insure proper wetting.

Cost information

- The operating cost (including pumping, oil makeup, heating and cooling) is \$0.05/kg-mol of recirculating oil.
- The system is to be operated for 8,000 hrs/annum.
- The installed cost, \$, of the absorption column (including auxiliaries, but excluding packing) is given by

$$\text{Installed cost of column} = 2,300\, H^{0.85}\, D^{0.95},$$ (2.36)

where H is the packing height (m) and D is the column diameter (m).
- The packing cost is \$800/m³.
- The oil-regeneration system is to be salvaged from a closing unit in the plant. Hence, its fixed cost will not be accounted for in the optimization calculations.

- The absorber and packing are assumed to depreciate linearly over five years with negligible salvage values.

Solution

At 300 K, the vapor pressure of benzene can be calculated from Eq. (2.35):

$$\ln p^o = 20.8 - \frac{2,789}{300 - 52}$$

i.e.,

$$p^o = 14,101 \text{ Pa},$$

Since the system is assumed to follow Raoult's law, then Eq. (2.3) can be used to give

$$m = \frac{14,101}{141,000}$$

$$\approx 0.1 \frac{mole\ fraction\ of\ benzene\ in\ air}{mole\ fraction\ of\ benzene\ in\ oil}.$$

As has been previously mentioned, the minimum TAC can be identified by iteratively varying ε. Since the inlet and outlet compositions of the rich stream as well as the inlet composition of the MSA are fixed, one can vary ε at the rich end of the exchanger (and consequently the outlet composition of the lean stream) to minimize the TAC of the system. In order to demonstrate this optimization procedure, let us first select a value of ε at the rich end of the exchanger equal to 1.5×10^{-3} and evaluate the system size and cost for this value.

Outlet composition of benzene in oil Let us set the outlet mole fraction of benzene in oil equal to its maximum practically feasible value given by Eq. (2.24), i.e.,

$$x^{out} = \frac{10^{-3}}{0.1} - 1.5 \times 10^{-3}$$

$$= 8.5 \times 10^{-3}.$$

Flowrate of oil A component material balance on benzene gives

$$L(8.5 \times 10^{-3} - 8 \times 10^{-4}) = 0.2(10^{-3} - 10^{-4})$$

or

$$L = 0.0234\ kg\ mol/s.$$

Operating cost

$$Annual\ operating\ \text{cost} = 0.05 \frac{\$}{kg\ mol\ oil} \times 0.0234 \frac{kg\ mol\ oil}{s} \times 3600 \times 8000 \frac{s}{yr}$$

$$= \$33,700/yr.$$

Column height According to Eq. (2.18b)

$$(y - y^*)_{\log mean} = \frac{(10^{-3} - 0.1 \times 8.5 \times 10^{-3})(10^{-4} - 0.1 \times 8 \times 10^{-4})}{\ln\left(\dfrac{10^{-3} - 0.1 \times 8.5 \times 10^{-3}}{10^{-4} - 0.1 \times 8 \times 10^{-4}}\right)}$$

$$= 6.45 \times 10^{-5}$$

Therefore, NTU_y can be calculated using Eq. (2.18a) as

$$NTU_y = \frac{10^{-3} - 10^{-4}}{6.45 \times 10^{-5}}$$

$$= 13.95,$$

and the height is obtained from Eq. (2.17a)

$$H = 0.6 \times 13.95$$

$$= 8.37 \text{ m}.$$

Column diameter

$$D = \sqrt{\frac{4(volumetric\ flowrate\ of\ gas)}{\pi(gas\ superficial\ velocity)}}$$

But

$$molar\ density\ of\ gas = \frac{P}{RT} \quad (see\ Appendix\ I)$$

$$= \frac{141}{8.3143 \times 300}$$

$$= 0.057\ kgmol/m^3.$$

Therefore,

$$Volumetric\ flowrate\ of\ gas = \frac{0.2\dfrac{kg\ mol}{s}}{0.057\dfrac{kg\ mol}{m^3}}$$

$$= 351\ m^3/s$$

and

$$D = \sqrt{\frac{4 \times 3.51}{3.14 \times 1.5}}$$

$$= 1.73\ m.$$

It is worth pointing out that the mass velocity of oil is

$$\frac{0.0234\frac{kg\ mole}{s} \times \frac{300\ kg}{kg\ mol}}{\frac{\pi}{4}(1.73)^2} \approx 3\ kg/s,$$

which is acceptable since it is greater than the minimum wetting velocity ($2.7\frac{kg}{m^2s}$).

Fixed cost

Fixed cost of installed shell and auxiliaries $= 2,300(8.37)^{0.85}(1.73)^{0.95}$
$$= \$23,600.$$
$$\text{Cost of packing} = (800)\frac{\pi}{4}(1.73)^2(8.37)$$
$$= \$15,700.$$

Total annualized cost

$$\text{TAC} = \text{annual operating cost} + \text{annualized fixed cost}$$
$$= 33,700 + \left(\frac{23,600 + 15,700}{5}\right)$$
$$= \$41,560/\text{yr}.$$

This procedure is carried out for various values of ε until the minimum TAC is identified. The results shown in Fig. 2.13 indicate that the value of $\varepsilon = 1.5 \times 10^{-3}$

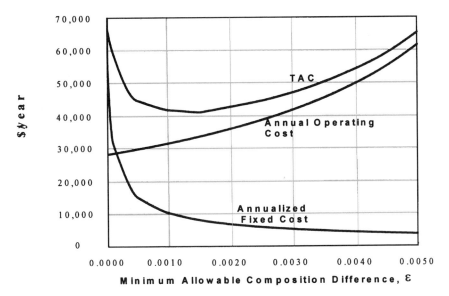

Figure 2.13 Using mass transfer driving force to trade off fixed cost versus operating cost.

used in the preceding calculations is the optimum one which corresponds to an optimum value of x^{out} being 0.0085. This solution has a minimum TAC of $41,560/yr.

Problems

2.1 A flue gas is to be desulfurized using water scrubbing in a stagewise column. It is desired to remove 95% of SO_2 from the flue gas containing 100 ppm SO_2. The flowrate of the gas is 0.03 kg mol/s (approximately 2 MMscfd). The water entering the absorber is free of SO_2. The equilibrium data for transferring SO_2 from the gas to the water may be described by:

$$y_i = 24x_j, \tag{2.37}$$

where y_i and x_j are the mole fractions of SO_2 in gas and water, respectively.

(a) What is the minimum flowrate of water needed to perform the desulfurization task? (Hint: set the minimum allowable composition difference at the rich end of the absorber equal to zero).

(b) If the flowrate of water is twice the minimum, what is the value of minimum allowable composition difference at the rich end of the absorber? How many theoretical stages are required?

2.2 A packed column is used in the desulfurization operation described in the previous problem. The overall height of transfer unit based on the gas phase is 0.7 m. When the flowrate of water is twice the minimum, what height of packing is needed?

2.3 A countercurrent moving-bed adsorption column is used to remove benzene from a gaseous emission. Activated carbon is employed as the adsorbent. The flowrate of the gas is 1.2 kg/s and it contains 0.027 wt/wt% of benzene. It is desired to recover 99% of this pollutant. The activated carbon entering the column has 2×10^{-4} wt/wt% of benzene. Over the operating range, the adsorption isotherm (Yaws et al., 1995) is linearized to

$$y_i = 0.0014x_j, \tag{2.38}$$

where y_i and x_j are the mass fractions of benzene in gas and activated carbon, respectively. The column height is 4 m and the value of HTU_y is 0.8 m. What should be the flowrate of activated carbon?

2.4 A multistage extraction column uses gas oil for the preliminary removal of phenol from wastewater. The flowrate of wastewater is 2.0 kg/s and its inlet mass fraction of phenol is 0.0358. The mass fraction of phenol in the wastewater exiting the column is 0.0168. Five kg/s of gas oil are used for extraction. The inlet mass fraction of phenol in gas oil is 0.0074. The equilibrium relation for the transfer of phenol from wastewater to gas oil is given by

$$y_i = 2.0x_j, \tag{2.39}$$

where y_i and x_j are the mass fractions of phenol in wastewater and gas oil, respectively. The overall column efficiency is 55%. How many actual stages are needed?

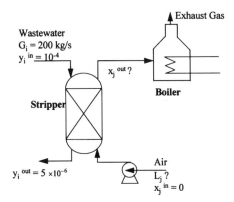

Figure 2.14 Stripping of TCE from wastewater.

2.5 Air stripping is used to remove 95% of the trichloroethylene (TCE, molecular weight = 131.4) dissolved in a 200 kg/s (3180 gpm) wastewater stream. The inlet composition of TCE in the wastewater is 100 ppm. Air (essentially free of TCE) is compressed to 202.6 kPa (2 atm) and diffused through a packed stripper. The TCE-laden air exiting the stripper is fed to the plant boiler, which burns almost all TCE. A schematic representation of the process is shown in Fig. 2.14.

The following physical and economic data are available:

Physical data

- The stripping operation takes place isothermally at 293 K and follows Henry's law.
- The equilibrium relation for stripping TCE from water is theoretically predicted using Eq. (2.5) to be

$$y_i = 0.0063x_j,$$

where y_i is the mass fraction of TCE in wastewater and x_i is the mass fraction of TCE in air (the value of Henry's coefficient corresponds to that reported in Mackay and Shiu (1981) as $1.065 \frac{gm\ mol}{kPa\ m^3}$).

- The air-to-water ratio is recommended by the packing manufacturer to be

$$25 \frac{m^3\ air}{m^3\ water}.$$

Stripper sizing criteria

- The maximum allowable superficial velocity of wastewater in the is taken as 0.02 m/s (approximately 30 gpm/ft^2; Cummins and Westrick, 1990).
- Overall height of transfer unit based on the liquid phase is given by

$$HTU_y = \text{Superficial velocity of wastewater}/K_y a$$

where K_y is the water-phase overall mass transfer coefficient and a is the surface area per unit volume of packing. The value of $K_y a$ is provided by the manufacturer to be 0.02 s^{-1}.

Cost information

- The operating cost for air compression is basically the electric utility needed for the isentropic compression. Electric energy needed to compress air may be calculated using Eq. (2.29). The isentropic efficiency of the compressor is taken as 60% and the electric energy cost is $0.06/kWhr.
- The system is to be operated for 8000 hr/annum.
- The fixed cost, $, of the stripper (including installation and auxiliaries, but excluding packing) is given by

$$\text{Fixed cost of column} = 4,700HD^{0.9}, \tag{2.41}$$

where H is the column height (m) and D is the column diameter (m).

- The cost of packing is $700/m^3$. $\tag{2.42}$
- The fixed cost the blower, $, is $12,000L_j^{0.6}$, $\tag{2.43}$

where L_j is the flowrate of air (kg/s).

- Assume negligible salvage value and a five-year linear depreciation.

(a) Estimate the column size, fixed cost and annual operating cost.

(b) Due to the potential error in the theoretically predicted value of Henry's coefficient, it is necessary to assess the sensitivity of your results to variations in the value of Henry's coefficient. Plot the column height, annualized fixed cost and annual operating cost versus α, the relative deviation from the nominal value of Henry's coefficient, for $0.5 \le \alpha \le 2.0$. The parameter α is defined by

$$\alpha = \text{Value of Henry's coefficient}/0.0063. \tag{2.44}$$

(c) Your company is planning to undertake extensive experimentation to obtain accurate values of Henry's coefficient that can be used in designing and evaluating the cost of this stripper. Based on your results, what would you recommend regarding the undertaking of these experiments?

2.6 Ammonia is to be absorbed from an air stream by contacting it with water in a packed column. The gas flowrate is 500 kg mol/hr and it contains 0.2 mol/mol% (2000 ppm) of ammonia. The molecular weight of the gas is 29, its temperature is 300 K and it has a pressure of 160 kPa. It is desired to reduce the ammonia content in the gas to 0.01 mol/mol% using the system shown in Fig. 2.15. Ammonia is first absorbed in water. The water is then fed to a wastewater treatment facility in which ammonia is used as a bio-nutrient.

What is the optimal flowrate of water that minimizes the TAC of the system?
The following data may be used.

Equilibrium data

- The absorption operation is assumed to be isothermal with the following equilibrium relation (King, 1980):

$$y_i = 1.41x_j, \tag{2.45}$$

where y_i and x_j are the mole fractions of ammonia in gas and water, respectively.

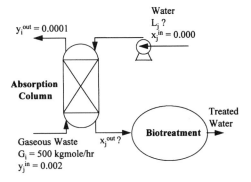

Figure 2.15 Ammonia absorption using water.

Absorber sizing criteria

- The overall-gas height of transfer unit for the packing is 0.8 m.

- The superficial velocity of the gas in the absorber is taken as 1.5 m/s to avoid flooding (Buonicore *et al.*, 1992).

- The manufacturer of the packing has recommended a range for the mass velocity of water between 2000 and $10000\,\frac{kg}{hr\cdot m^2}$.

Cost information

- The operating cost of the water (including pumping and wastewater treatment is $0.005/kg mole water.

- The system is to be operated for 8,000 hr/annum.

- The purchased cost, $, of the absorption column (fiberglass reinforced plastic shell and auxiliaries excluding packing) is given by (Vatavuk, 1995):

$$\text{Purchased cost of column(\$)} = 1,313\,S. \qquad (2.46)$$

S is the absorber surface area (m^2) which can be calculated from

$$S = \pi DH + \frac{\pi}{2}D^2, \qquad (2.47)$$

where D is the column diameter (m) and H is the column height (m) taken as 1.2 times the packing height.

 The packing cost is $600/m^3.

- The ratio of the fixed capital investment to the purchased equipment cost (Lang factor) is taken as 4.83.

- The absorber and packing are assumed to depreciate linearly over three years with negligible salvage values.

References

Astarita, G., Savage, D. W., and Bisio, A. (1983), "Gas Treating with Chemical Solvents," Wiley, New York.

Cheremisinoff, P. N. and Ellerbusch, F. (1980). "Carbon Adsorption Handbook." Ann Arbor Sci. Publ., Ann Arbor, MI.

Clark, R. M. (1990). Unit process research for removing volatile organic chemicals from drinking water: An overview. In "Significance and Treatment of Volatile Organic Compounds in Water Supplies," (N. M. Ram, R. F. Christman, and K. P. Cantor, eds.), Lewis Publishers, Chelsea, MI.

Crowl, D. A. and Louvar, J. F. (1990). "Chemical Process Safety," p. 161. Prentice Hall, Engelwood Cliffs, NJ.

Cummins, M. D. and Westrick, J. W. (1990). Treatment technologies and costs for removing VOC's from water. In "Significance and Treatment of Volatile Organic Compounds in Water Supplies," (N. M. Ram, R. F. Christman, and K. P. Cantor, eds.), p. 277. Lewis Publishers, Chelsea, MI.

Fleming, J. L. (1989). "Volatilization Technologies for Removing Organics from Water." Noyes Data Corp, Park Ridge, NJ.

Francis, A. W. (1963). "Liquid-Liquid Equilibrium." Interscience, New York.

Geankoplis, C. J. (1983). "Transport Processes and Unit Operations," 2nd ed. Allyn and Bacon, Boston.

Henley, E. J. and Seader, J. D. (1981). "Equilibrium-Stage Separation Operations in Chemical Engineering." Wiley, New York.

Hwang, Y., Olson, J. D., and Keller, G. E. II (1992). Steam Stripping for Removal of Organic Pollutants from Water: Parts 1 and 2. *Ind. Eng. Chem. Res.*, **31**(7), 1753–1768.

King, C. J. (1980). "Separation Processes," 2nd ed., McGraw Hill, New York.

Kohl, A., and Riesenfeld, F. (1985). "Gas Purification." 4th ed., Gulf Publ. Co, Houston, TX.

Kremser, A. (1930). *Nat. Pet. News*, **22**(21), 42–43.

Lo, T. C., Baird, M. H. I., and Hanson, C. (1983). "Handbook of Solvent Extraction." Wiley-Interscience, New York.

Mackay, D. and Shiu, W. Y. (1981). A critical review of Henry's law constants for chemicals of environmental interest. *J. Phys. Chem. Ref. Data*, **10**(4), 1175–1199.

McCabe, W. L., Smith, J. C., and Harriot, P. (1993). "Unit Operations of Chemical Engineering," 5th ed., McGraw Hill, New York.

Perrich, J. R. (1981). "Activated Carbon Adsorption for Wastewater Treatment." CRC Press, Boca Raton, FL.

Perry, R. H. and Green, D. (1984). "Perry's Chemical Engineers' Handbook," 6th ed., McGraw Hill, New York.

Peters, M. S. and Timmerhaus, K. D. (1991). "Plant Design and Economics for Chemical Engineers," 4th ed., McGraw Hill, New York.

Reid, R. C., Prausnitz, J. M., and Poling, B. E. (1987). "The Properties of Gases and Liquids," 4th ed. McGraw Hill, New York.

Stenzel, M. H. (1993). Remove organics by activated carbon adsorption. *Chem. Eng. Prog.*, **89**(4), 36–43.

Stenzel, M. H. and Merz, W. J. (1989). Use of carbon adsorption processes in groundwater treatment. *Environ. Prog.*, **8**(4), 257–264.

Treybal, R. E. (1980). "Mass Transfer Operations," 2nd ed., McGraw Hill, New York.

U.S. Environmental Protection Agency (1980). "Carbon Adsorption Isotherms for Toxic Organics," EPA-600/8-80-023. USEPA, Cincinnati, OH.

Valenzuela, D. P. and Myers, A. (1989). "Adsorption Equilibrium Data Handbook." Prentice Hall, Englewood Cliffs, NJ.

Vatavuk, W. M. (1995). A potpourri of equipment prices. *Chem. Eng.*, pp. 68–73.

Wankat, P. C. (1994). "Equilibrium Staged Separations." Elsevier, New York.

Yang, R. T. (1987). "Gas Separation by Adsorption Processes." Butterworth, Boston.

Yaws, C. L. (1992). "Thermodynamic and Physical Property Data" Gulf Publ. Co Houston, TX.

Yaws, C. L., Bu, L., and Nijhawan, S. (1995). Adsorption capacity data for 283 organic compounds. *Environ. Eng. World*, May–June, pp. 16–19.

CHAPTER THREE

Synthesis of Mass-Exchange Networks

A Graphical Approach

3.1 A Network versus a Unit

Chapter Two has provided an overview of the principles of modeling, designing, and operating individual mass exchange units. However, in many industrial situations the problem of selecting, designing, and operating a mass-exchange system should not be confined to assessing the performance of individual mass exchangers. In such situations, there are several rich streams (sources) from which mass has to be removed, and many mass separating agents can be used for removing the targeted species. Therefore, adopting a systemic network approach can provide significant technical and economic benefits. In this approach, a mass-exchange system is selected, and designed by *simultaneously* screening all candidate mass exchange operations to identify the optimum system. This chapter defines the problem of synthesizing mass-exchange networks (MENs) discusses its challenging aspects and provides a graphical approach for the synthesis of MENs.

3.2 Problem Scope, Significance, and Complexity

As has been discussed in Chapter Two, a mass exchanger is any direct-contact countercurrent mass-transfer unit that employs an MSA. The realm of mass-exchange operations includes absorption, adsorption, ion exchange, solvent extraction, leaching, and stripping. Examples of MSAs are solvents, adsorbents, ion-exchange resins and stripping agents. The synthesis and analysis of a waste-recovery system that employs multiple mass-exchange operations is not a straightforward task. In a typical industrial situation, many MSAs are candidates for removing the undesirable species. Given the relatively small concentrations involved in most environmental applications, one should consider only those MSAs that are thermodynamically feasible. In addition, the designer has to select MSA types and flowrates that minimize cost. Furthermore, the mass exchangers should be properly interconnected and sized to yield a cost-effective solution to the given waste management task.

Motivated by the need to simultaneously screen candidate MSAs while incorporating thermodynamic and economic considerations, El-Halwagi and Manousiouthakis (1989) introduced the powerful notion of synthesizing MENs and developed systematic techniques for their optimal design. At present, there is a comprehensive methodology for the design of MENs that can tackle a wide variety of environmental applications (El-Halwagi and Manousiouthakis, 1990a, b; El-Halwagi and El-Halwagi, 1992; El-Halwagi and Srinivas, 1992; El-Halwagi et al., 1992; Allen et al., 1992; Dunn and El-Halwagi, 1993; Gupta and Manousiouthakis, 1993; El-Halwagi, 1993; Wang and Smith, 1994; Srinivas and El-Halwagi, 1994a, b; Papalexandri and Pistikopoulos, 1994; Stanley and El-Halwagi, 1995; Samdani, 1995; Huang and Edgar, 1995; Huang and Fan, 1995; Garrison et al., 1995; Warren et al., 1995; Kiperstok and Sharratt, 1995; Zhu and El-Halwagi, 1995; Dhole et al., 1996). The following section defines the basic form of the MEN synthesis task.

3.3 Mass-Exchange Network Synthesis Task

The problem of synthesizing MENs can be stated as follows: Given a number N_R of waste (rich) streams (sources) and a number N_S of MSAs (lean streams), it is desired to synthesize a cost-effective network of mass exchangers that can preferentially transfer certain undesirable species from the waste streams to the MSAs. Given also are the flowrate of each waste stream, G_i, its supply (inlet) composition y_i^s, and its target (outlet) composition y_i^t, where $i = 1, 2, \ldots, N_R$. In addition, the supply and target compositions, x_j^s and x_j^t, are given for each MSA, where $j = 1, 2, \ldots, N_S$. The flowrate of each MSA is unknown and is to be determined so as to minimize the network cost. Figure 3.1 is a schematic representation of the problem statement.

The candidate lean streams can be classified into N_{SP} process MSAs and N_{SE} external MSAs (where $N_{SP} + N_{SE} = N_S$). The process MSAs already exist on plant site and can be used for the removal of the undesirable species at a very low cost (virtually free). The flowrate of each process MSA that can be used for mass exchange is bounded by its availability in the plant, i.e.,

$$L_j \leq L_j^c \quad j = 1, 2, \ldots, N_{SP}, \tag{3.1}$$

where L_j^c is the flowrate of the jth MSA that is available in the plant. On the other hand, the external MSAs can be purchased from the market. Their flowrates are to be determined according to the overall economic considerations of the MEN.

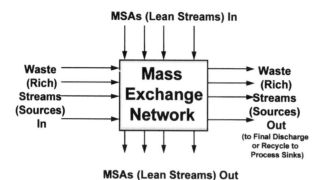

Figure 3.1 Schematic representation of the MEN synthesis problem.

Typically, waste streams are either disposed of or forwarded to process sinks (equipment) for recycle/reuse. In case of final discharge, the target composition of the undesirable species in each waste stream corresponds to the environmental regulations. On the other hand, if the intercepted waste stream is to be recycled to a process sink, its target composition should satisfy the constraints imposed by the process sink.

The target composition of the undesirable species in each MSA is assigned by the designer based on the specific circumstances of the application. The nature of such circumstances may be physical (e.g., maximum solubility of the pollutant in the MSA), technical (e.g., to avoid excessive corrosion, viscosity or fouling), environmental (e.g., to comply with environmental regulations), safety (e.g., to stay away from flammability limits), or economic (e.g., to optimize the cost of subsequent regeneration of the MSA).

As can be inferred from the problem statement, for a given waste-reduction situation the MEN synthesis task attempts to provide cost-effective solutions to the following design questions:

- Which mass-exchange operations should be used (e.g., absorption, adsorption)?
- Which MSAs should be selected (e.g., which solvents, adsorbents)?
- What is the optimal flowrate of each MSA?
- How should these MSAs be matched with the waste streams (i.e., stream pairings)?
- What is the optimal system configuration (e.g., how should these mass exchangers be arranged? Is there any stream splitting and mixing?)?

The foregoing design questions are highly combinatorial so any exhaustive enumeration technique would be hopelessly complicated. A hit-and-miss or trial-and-error approach to conjecturing the solution is likely to fail because it cannot consider the overwhelming number of decisions to be made. Hence, the designer needs practical tools to systematically address the MEN synthesis task. These tools are provided in the rest of this chapter along with Chapters Four through Six. They are based on a useful process synthesis technique that is commonly referred to as the "targeting approach."

3.4 The Targeting Approach

The targeting approach is based on the identification of performance targets ahead of design and without prior commitment to the final network configuration. In the context of synthesizing MENs, two useful targets can be established:

1. **Minimum cost of MSAs.** By integrating the thermodynamic aspects of the problem with the cost data of the MSAs, one can indeed identify the minimum cost of MSAs and the flowrate of each MSA required to undertake the assigned mass-exchange duty. This can be accomplished without actually designing the network. Since the cost of MSAs is typically the dominant operating expense, this target is aimed at minimizing the operating cost of the MEN. Any design featuring the minimum cost of MSAs will be referred to as a minimum operating cost (MOC) solution.

2. **Minimum number of mass exchanger units.** Combinatorics determines the minimum number of mass exchanger units required in the network. This objective attempts to minimize indirectly the fixed cost of the network, since the cost of each mass exchanger is usually a concave function of the unit size. Furthermore, in a practical context it is desirable to minimize the number of separators so as to reduce pipework, foundations, maintenance, and instrumentation. Normally, the minimum number of units is related to the total number of streams by the following expression (El-Halwagi and Manousiouthakis, 1989).

$$U = N_R + N_S - N_i \tag{3.2}$$

where N_i is the number of independent synthesis subproblems into which the original synthesis problem can be subdivided. In most cases, there is only one independent synthesis subproblem.

In general, these two targets are incompatible. Later, systematic techniques will be presented to enable the identifying an MOC solution and then minimizing the number of exchangers satisfying the MOC.

3.5 The Corresponding Composition Scales

A particularly useful concept in synthesizing MENs is the notion of "corresponding composition scales." It is a tool for incorporating thermodynamic constraints

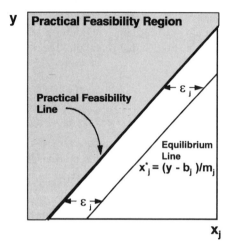

Figure 3.2 Establishing corresponding composition scales.

of mass exchange by establishing a one-to-one correspondence among the compositions of all streams for which mass transfer is thermodynamically feasible. This concept is based on a generalization of the notion of a "minimum allowable composition difference" presented in Section 2.5. To demonstrate this concept, let us consider a mass exchanger for which the equilibrium relation governing the transfer of the pollutant from the waste stream, i, to the MSA, j, is given by the linear expression

$$y_i = m_j x_j^* + b_j, \tag{3.3}$$

which indicates that for a waste stream composition of y_i, the maximum theoretically attainable composition of the MSA is x_j^*. By employing a minimum allowable composition difference of ε_j, one can draw a "practical-feasibility line" that is parallel to the equilibrium line but offset to its left by a distance ε_j (Fig. 3.2). In order for an operating line to be practically feasible, it must lie in the region to the left of the practical-feasibility line. Hence, for any pair (y_i, x_j) lying on the practical-feasibility line, two statements can be made. For a given y_i, the value x_j corresponds to the maximum composition of the pollutant that is practically achievable in the MSA. Conversely, for a given x_j, the value y_i corresponds to the minimum composition of the pollutant in the waste stream that is needed to practically transfer the pollutant from the waste stream to the MSA.

It is important to derive the mathematical expression relating y_i and x_j on the practical-feasibility line. For a given y_i, the values of x_j can be obtained by evaluating x_j^* that is in equilibrium with y_i then subtracting ε_j, i.e.,

$$x_j = x_j^* - \varepsilon_j \tag{3.4a}$$

or

$$x_j^* = x_j + \varepsilon_j. \tag{3.4b}$$

Substituting from (3.4b) into (3.3), one obtains

$$y_i = m_j(x_j + \varepsilon_j) + b_j \tag{3.5a}$$

or

$$x_j = \frac{y_i - b_j}{m_j} - \varepsilon_j. \tag{3.5b}$$

Equation (3.5) can be used to establish a one-to-one correspondence among all composition scales for which mass exchange is feasible. Since most environmental applications involve dilute systems, one can assume that these systems behave ideally. Hence, the transfer of the pollutant is indifferent to the existence of other species in the waste stream. In other words, even if two waste streams contain species that are not identical, but share the same composition of a particular pollutant, the equilibrium composition of the pollutant in an MSA will be the same for both waste streams. Hence, a single composition scale, y, can be used to represent the concentration of the pollutant in any waste stream. Next, (3.5) can be employed to generate N_S scales for the MSAs. For a given set of corresponding composition scales $\{y, x_1, x_2, \ldots, x_j, \ldots, x_{NS}\}$ it is thermodynamically and practically feasible to transfer the pollutant from any waste stream to any MSA. In addition, it is also feasible to transfer the pollutant from any waste stream of a composition y_i to any MSA which has a composition less than the x_j obtained from (3.5b).

3.6 The Pinch Diagram

In order to minimize the cost of MSAs, it is necessary to make maximum use of process MSAs before considering the application of external MSAs. In assessing the applicability of the process MSAs to remove the pollutant, one must consider the thermodynamic limitations of mass exchange. Towards this end, one may use a graphical approach referred to as the "pinch diagram" (El-Halwagi and Manousiouthakis, 1989). The initial step in constructing the pinch diagram is creating a global representation for all the rich streams. This representation is accomplished by first plotting the mass exchanged by each rich stream versus its composition. Hence, each rich stream is represented as an arrow whose tail corresponds to its supply composition and its head to its target composition. The slope of each arrow is equal to the stream flowrate. The vertical distance between the tail and the head of each arrow represents the mass of pollutant that is lost by

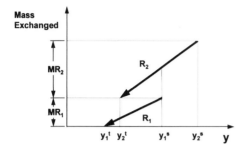

Figure 3.3 Representation of mass exchanged by two rich streams.

that rich stream according to

Mass of pollutant lost from the ith rich stream

$$MR_i = G_i\left(y_i^s - y_i^t\right), \quad i = 1, 2, \ldots, N_R. \tag{3.6}$$

It is worth noting that the vertical scale is only relative. Any stream can be moved up or down while preserving the same vertical distance between the arrow head and tail and maintaining the same supply and target compositions. A convenient way of vertically placing each arrow is to stack the waste streams on top of one another, starting with the waste stream having the lowest target composition (see Fig. 3.3 for an illustration for two waste streams).

Having represented the individual rich streams, we are now in a position to construct the rich composite stream. A rich composite stream represents the cumulative mass of the pollutant lost by all the rich streams. It can be readily obtained by using the "diagonal rule" for superposition to add up mass in the overlapped regions of streams. Hence, the rich composite stream is obtained by applying linear superposition to all the rich streams. Figure 3.4 illustrates this concept for two rich streams.

Next, a global representation of all process lean streams is developed as a lean composite stream. First, we establish N_{SP} lean composition scales (one for each process MSA) that are in one-to-one correspondence with the rich scale according to the method outlined in Section 3.5. Next, the mass of pollutant that can be gained by each process MSA is plotted versus the composition scale of that MSA. Hence, each process MSA is represented as an arrow extending between supply and target compositions (see Fig. 3.5 for a two-MSA example). The vertical distance between the arrow head and tail is given by

Mass of pollutant that can be gained by the jth process MSA

$$MS_j = L_j^c\left(x_j^t - x_j^s\right) \quad j = 1, 2, \ldots, N_{SP}. \tag{3.7}$$

Once again, the vertical scale is only relative and any stream can be moved up or down on the diagram. A convenient way of vertically placing each arrow is to

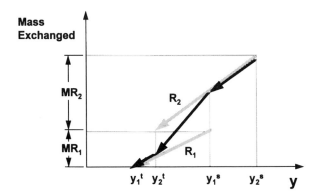

Figure 3.4 Constructing a rich composite stream using superposition.

stack the process MSAs on top of one another starting with the MSA having the lowest supply composition (Fig. 3.6). Hence, a lean composite stream representing the cumulative mass of the pollutant gained by all the MSAs is obtained by using the diagonal rule for superposition.

Next, both composite streams are plotted on the same diagram (Fig. 3.7). On this diagram, thermodynamic feasibility of mass exchange is guaranteed when the lean composite stream is always above the waste composite stream. This is equivalent to ensuring that at any mass-exchange level (which corresponds to a horizontal line), the composition of the lean composite stream is located to the left of the waste composite stream, asserting thermodynamic feasibility. Therefore,

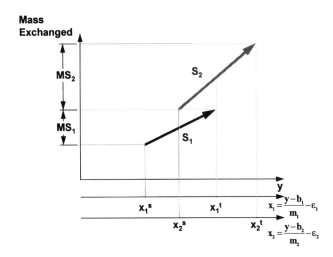

Figure 3.5 Representation of mass exchanged by two process MSAs.

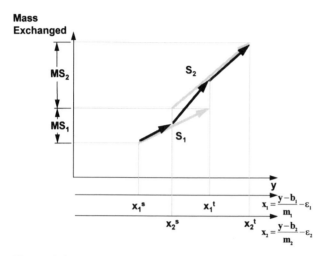

Figure 3.6 Construction of the lean composite stream using superposition.

the lean composite stream can be slid down until it touches the waste composite stream. The point where the two composite streams touch is called the "mass-exchange pinch point"; hence the name "pinch diagram" (Fig. 3.7).

On the pinch diagram, the vertical overlap between the two composite streams represents the maximum amount of the pollutant that can be transferred from the waste streams to the process MSAs. It is referred to as the "integrated mass

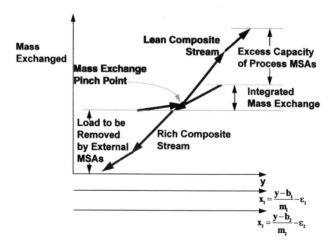

Figure 3.7 The mass-exchange pinch diagram (from El-Halwagi and Manousiouthakis, 1989. Reproduced with permission from the American Institute of Chemical Engineers. Copyright © 1989. All rights reserved).

exchange." The vertical distance of the lean composite stream which lies above the upper end of the waste composite stream is referred to as "excess process MSAs." It corresponds to the capacity of the process MSAs to remove pollutants that cannot be used because of thermodynamic infeasibility. According to the designer's preference or to the specific circumstances of the process such excess can be eliminated from service by lowering the flowrate and/or the outlet composition of one or more of the process MSAs. Finally, the vertical distance of the waste composite stream which lies below the lower end of the lean composite stream corresponds to the mass of pollutant to be removed by external MSAs.

As can be seen from Fig. 3.7, the pinch decomposes the synthesis problem into two regions: a rich end and a lean end. The rich end comprises all streams or parts of streams richer than the pinch composition. Similarly, the lean end includes all the streams or parts of streams leaner than the pinch composition. Above the pinch, exchange between the rich and the lean process streams takes place. External MSAs are not required. Using an external MSA above the pinch will incur a penalty of eliminating an equivalent amount of process lean streams from service. On the other hand, below the pinch, both the process and the external lean streams should be used. Furthermore, Fig. 3.7 indicates that if any mass is transferred across the pinch, the composite lean stream will move upward and, consequently, external MSAs in excess of the minimum requirement will be used. Therefore, to minimize the cost of external MSAs, mass should not be transferred across the pinch. It is worth pointing out that these observations are valid only for the class of MEN problems covered in this chapter. When the assumptions employed in this chapter are relaxed, more general conclusions can be made. For instance, it will be shown later that the pinch analysis can still be undertaken even when there are no process MSAs in the plant. The pinch characteristics will be generalized in Chapters Five and Six.

Example 3.1: Recovery of Benzene from Gaseous Emission of a Polymer Production Facility

Figure 3.8 shows a simplified flowsheet of a copolymerization plant. The copolymer is produced via a two-stage reaction. The monomers are first dissolved in a benzene-based solvent. The mixed-monomer mixture is fed to the first stage of reaction where a catalytic solution is added. Several additives (extending oil, inhibitors, and special additives) are mixed in a mechanically stirred column. The resulting solution is fed to the second-stage reactor, where the copolymer properties are adjusted. The stream leaving the second-stage reactor is passed to a separation system which produces four fractions: copolymer, unreacted monomers, benzene, and gaseous waste. The copolymer is fed to a coagulation and finishing section. The unreacted monomers are recycled to the first-stage reactor, and the recovered

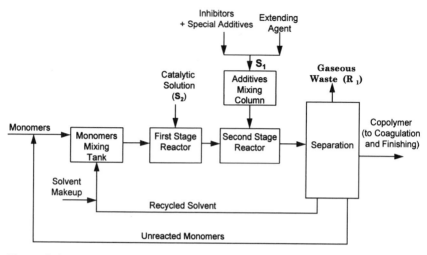

Figure 3.8 A simplified flowsheet of a copolymerization process.

benzene is returned to the monomer-mixing tank. The gaseous waste, R_1, contains benzene as the primary pollutant that should be recovered. The stream data for R_1 are given in Table 3.1.

Two process MSAs and one external MSA are considered for recovering benzene from the gaseous waste. The two process MSAs are the additives, S_1, and the liquid catalytic solution, S_2. They can be used for benzene recovery at virtually no operating cost. In addition to its positive environmental impact, the recovery of benzene by these two MSAs offers an economic incentive since it reduces the benzene makeup needed to compensate for the processing losses. Furthermore, the additives mixing column can be used as an absorption column by bubbling the gaseous waste into the additives. The mixing pattern and speed of the mechanical stirrer can be adjusted to achieve a wide variety of mass-transfer tasks. The stream data for S_1 and S_2 are given in Table 3.2. The equilibrium data for benzene in the

Table 3.1
Data of Waste Stream for the Benzene Removal Example

Stream	Description	Flowrate G_i, kg mol/s	Supply composition (mole fraction) y_i^s	Target composition (mole fraction) y_i^t
R_1	Off-gas from product separation	0.2	0.0020	0.0001

Table 3.2
Data of Process Lean Streams for the Benzene Removal Example

Stream	Description	Upper bound on flowrate L_j^C kg mol/s	Supply composition of benzene (mole fraction) x_j^s	Target composition of benzene (mole fraction) x_j^t
S_1	Additives	0.08	0.003	0.006
S_2	Catalytic solution	0.05	0.002	0.004

two process MSAs are given by:

$$y = 0.25x_1 \tag{3.8}$$

and

$$y = 0.50x_2, \tag{3.9}$$

where y, x_1 and x_2 are the mole fractions of benzene in the gaseous waste, S_1 and S_2 respectively. For control purposes, the minimum allowable composition difference for S_1 and S_2 should not be less than 0.001.

The external MSA, S_3, is an organic oil that can be regenerated using flash separation. The operating cost of the oil (including pumping, makeup, and regeneration) is $0.05/kg mol of recirculating oil. The equilibrium relation for transferring benzene from the gaseous waste to the oil is given by

$$y = 0.10x_3. \tag{3.10}$$

The data for S_3 are given in Table 3.3. The absorber sizing equations and fixed cost were given in Example 2.2. Using the graphical pinch approach, synthesize a cost-effective MEN that can be used to remove benzene from the gaseous waste (Fig. 3.9).

Table 3.3
Data for the External MSA for the Benzene Removal Example

Stream	Description	Upper bound on flowrate L_j^C kg mole/s	Supply composition of benzene (mole fraction) x_j^s	Target composition of benzene (mole fraction) x_j^t
S_3	Organic oil	∞	0.0008	0.0100

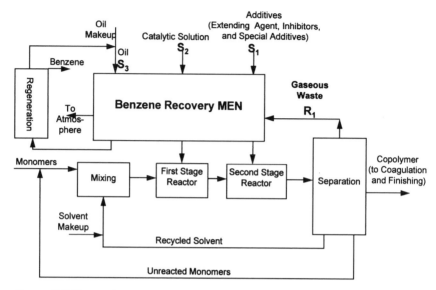

Figure 3.9 The copolymerization process with a benzene recovery MEN.

Solution

Constructing the pinch diagram As has been described earlier, the waste composite stream is first plotted as shown in Fig. 3.10. Next, the lean composite stream is constructed for the two process MSAs. Equation (3.5) is employed to generate the correspondence among the composition scales y, x_1 and x_2. The least permissible values of the minimum allowable composition difference are used ($\varepsilon_1 = \varepsilon_2 = 0.001$). Later, it will be shown that these values are optimum for a MOC solution. Next, the mass exchangeable by each of the two process lean streams is represented as an arrow versus its respective composition scale (Fig. 3.11a). As demonstrated by Fig. 3.11b, the lean composite stream is obtained by applying superposition to the two lean arrows. Finally, the pinch diagram is constructed by combining Figs. 3.10 and 3.11b. The lean composite stream is slid vertically until it is completely above the rich composite stream.

Interpreting results of the pinch diagram As can be seen from Fig. 3.12, the pinch is located at the corresponding mole fractions $(y, x_1, x_2) = (0.0010, 0.0030, 0.0010)$. The excess capacity of the process MSAs is 1.4×10^{-4} kg mol benzene/s and cannot be used because of thermodynamic and practical-feasibility limitations. This excess can be eliminated by reducing the outlet compositions and/or flowrates of the process MSAs. Since the inlet composition of S_2 corresponds to a mole fraction of 0.0015 on the y scale, the waste load immediately

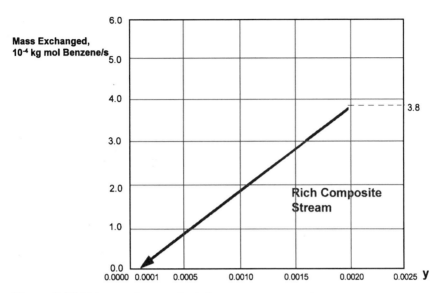

Figure 3.10 Rich composite stream for the benzene recovery example.

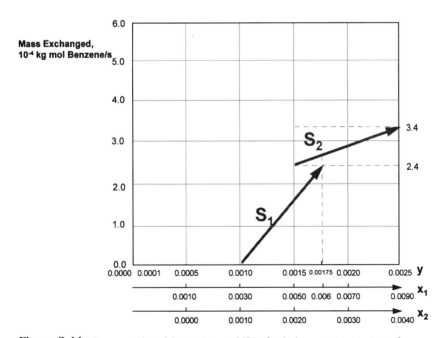

Figure 3.11a Representation of the two process MSAs for the benzene recovery example.

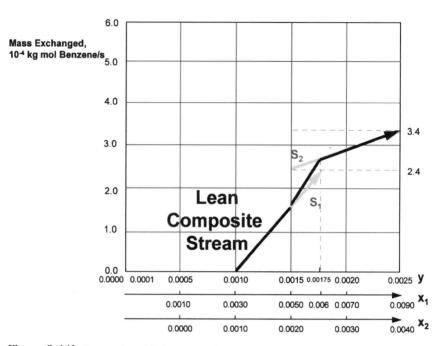

Figure 3.11b Construction of the lean composite stream for the two process MSA's of the benzene recovery example.

above the pinch (from $y = 0.0010$ to $y = 0.0015$) cannot be removed by S_2. Therefore, S_1 must be included in a MOC solution. Indeed, S_1 alone can be used to remove all the waste load above the pinch (2×10^{-4} kg mol benzene/s). To reduce the fixed cost by minimizing the number of mass exchangers, it is preferable to use a single solvent above the pinch rather than two solvents. This is particularly attractive given the availability of the mechanically stirred additives-mixing column for absorption. Hence, the excess capacity of the process MSAs is eliminated by avoiding the use of S_2 and reducing the flowrate and/or outlet composition of S_1. There are infinite combinations of L_1 and x_1^{out} that can be used to remove the excess capacity of S_1 according to the following material balance:

Benzene load above the pinch to be removed by $S_1 = L_1\left(x_1^{out} - x_1^s\right)$,   (3.11a)

i.e.,

$$2 \times 10^{-4} = L_1\left(x_1^{out} - 0.003\right). \qquad (3.11b)$$

Nonetheless, since the additives-mixing column will be used for absorption, the whole flowrate of S_1 (0.08 kg mol/s) should be fed to the column. Hence, according to Eq. (3.11b), the outlet composition of S_1 is 0.0055. The same result can be obtained graphically as shown in Fig. 3.13. It is worth recalling that the target

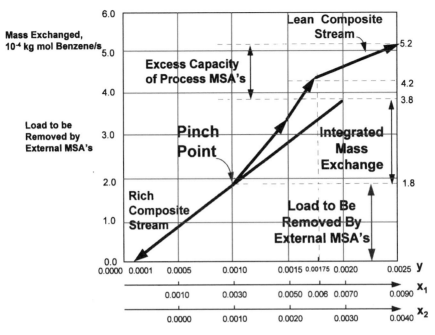

Figure 3.12 The pinch diagram for the benzene recovery example ($\varepsilon_1 = \varepsilon_2 = 0.001$).

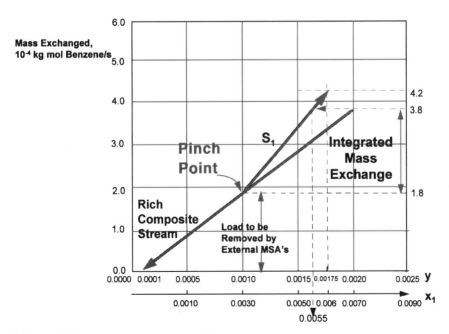

Figure 3.13 Graphical identification of x_1^{out}.

composition of an MSA is only an upper bound on the actual value of the outlet composition. As was shown in this example, the outlet composition of an MSA is typically selected so as to optimize the cost of the system.

Selection of the optimal value of ε_1 Since S_1 is a process MSA with almost no operating cost and since it is to be used in process equipment (the mechanically stirred column) that does not require additional capital investment for utilization as an absorption column, S_1 should be utilized to its maximum practically feasible capacity for absorbing benzene. The remaining benzene load (below the pinch) is to be removed using the external MSA. The higher the benzene load below the pinch, the higher the operating and fixed costs. Therefore, in this example it is desired to maximize the integrated mass exchanged above the pinch. As can be seen on the pinch diagram when ε_1 increases, the x_1 axis moves to the right relative to the y axis and, consequently, the extent of integrated mass exchange decreases leading to a higher cost of external MSAs. For instance, Fig. 3.14 demonstrates the pinch diagram when ε_1 is increased to 0.002. The increase of ε_1 to 0.002 results in a load of 2.3×10^{-4} kg mol benzene/s to be removed by external MSAs (compared to 1.8×10^{-4} kg mol benzene/s for $\varepsilon_1 = 0.001$), an integrated mass exchange of 1.5×10^{-4} kg mol benzene/s (compared to 2.0×10^{-4} kg mol

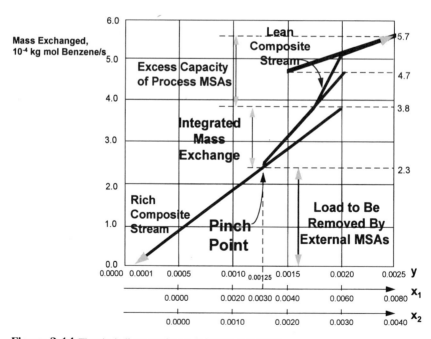

Figure 3.14 The pinch diagram when ε_1 is increased to 0.002.

benzene/s for $\varepsilon_1 = 0.001$) and an excess capacity of process MSAs of 1.9×10^{-4} kg mol benzene/s (compared to 1.4×10^{-4} kg mol benzene/s for $\varepsilon_1 = 0.001$). Thus; the optimum ε_1 in this example is the smallest permissible value given in the problem statement to be 0.001.

It is worth noting that there is no need to optimize over ε_2. As previously shown, when ε_2 was set equal to its lowest permissible value (0.001), S_1 was selected as the optimal process MSA above the pinch. On the pinch diagram, as ε_2 increases, S_2 moves to the right, and the same arguments for selecting S_1 over S_2 remain valid.

Optimizing the use of the external MSA The pinch diagram (Fig. 3.12) demonstrates that below the pinch, the load of the waste stream has to be removed by the external MSA, S_3. This renders the remainder of this example identical to Example 2.2. Therefore, the optimal flowrate of S_3 is 0.0234 kg mol/s and the optimal outlet composition of S_3 is 0.0085. Furthermore, the minimum total annualized cost of the benzene recovery system is \$41,560/yr (see Fig. 2.13).

Constructing the synthesized network The previous analysis shows that the MEN comprises two units: one above the pinch in which R_1 is matched with S_1, and one below the pinch in which the remainder load of R_1 is removed using S_3. Figure 3.15 illustrates the network configuration.

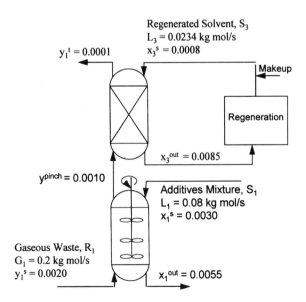

Figure 3.15 Optimal MEN for the benzene recovery example.

Unlike this example, some cases involve the construction of less intuitively apparent MEN configurations. Chapters Five and Six provide systematic rules for matching streams and configuring the network. Furthermore, many MEN-synthesis problems require the screening of multiple external MSAs. This issue is addressed by the next example.

Example 3.2: Dephenolization of Aqueous Wastes

Consider the oil-recycling plant shown in Fig. 3.16. In this plant, two types of waste oil are handled: gas oil and lube oil. The two streams are first deashed and demetallized. Next, atmospheric distillation is used to obtain light gases, gas oil, and a heavy product. The heavy product is distilled under vacuum to yield lube oil. Both the gas oil and the lube oil should be further processed to attain desired properties. The gas oil is steam stripped to remove light and sulfur impurities, then hydrotreated. The lube oil is dewaxed/deasphalted using solvent extraction followed by steam stripping.

The process has two main sources of waste water. These are the condensate streams from the steam strippers. The principal pollutant in both wastewater streams is phenol. Phenol is of concern primarily because of its toxicity, oxygen depletion, and turbidity. In addition, phenol can cause objectionable taste and odor in fish flesh and potable water.

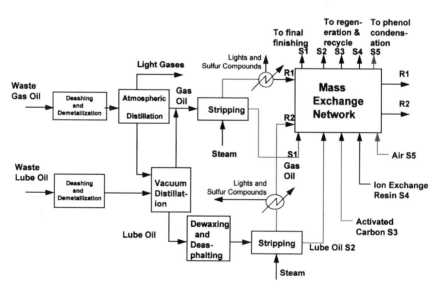

Figure 3.16 Schematic representation of an oil recycling plant.

Table 3.4
Data of Waste Streams for the Dephenolization Example

Stream	Description	Flowrate G_i, kg/s	Supply composition y_i^s	Target composition y_i^t
R_1	Condensate from first stripper	2	0.050	0.010
R_2	Condensate from second stripper	1	0.030	0.006

Several techniques can be used to separate phenol. Solvent extraction using gas oil or lube oil (process MSAs: S_1 and S_2, respectively) is a potential option. Besides the purification of wastewater, the transfer of phenol to gas oil and lube oil is a useful process for the oils. Phenol tends to act as an oxidation inhibitor and serves to improve color stability and reduce sediment formation. The data for the waste streams and the process MSAs are given in Tables 3.4 and 3.5, respectively.

Three external technologies are also considered for the removal of phenol. These processes include adsorption using activated carbon, S_3, ion exchange using a polymeric resin, S_4, and stripping using air, S_5. The equilibrium data for the transfer of phenol to the jth lean stream is given by $y = m_j x_j$ where the values of m_j are 2.00, 1.53, 0.02, 0.09 and 0.04 for S_1, S_2, S_3, S_4 and S_5, respectively. Throughout this example, a minimum allowable composition difference, ε_j, of 0.001(kg phenol)/(kg MSA) will be used.

The pinch diagrams The pinch diagram is constructed as shown in Fig. 3.17. The pinch is located at the corresponding mass fraction (y, x_1, x_2) = (0.0168, 0.0074, 0.0100). The excess capacity of the process MSAs is 0.0184 kg phenol/s.

Table 3.5
Data of Process MSAs for the Dephenolization Example

Stream	Description	Upper bound on flowrate L_j^c, kg/s	Supply composition, x_j^s	Target composition, x_j^t
S_1	Gas oil	5	0.005	0.015
S_2	Lube oil	3	0.010	0.030

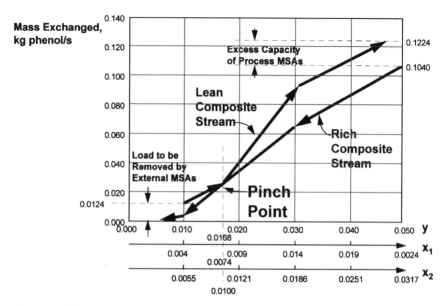

Figure 3.17 The pinch diagram for the dephenolization example.

This excess can be eliminated by reducing the outlet compositions and/or flowrates of the process MSAs. For instance, if the designer elects to remove this excess by lowering the flowrate of S_2, the actual flowrate of S_2 can then be calculated as follows

$$L_2 = 3 - \frac{0.0184}{0.03 - 0.01}$$
$$= 2.08 \ kg/s.$$

(3.12)

Figure 3.17 also indicates that 0.0124 kg phenol/s are to be removed using external MSAs. The following section discusses screening the three MSAs to select the one that yields a MOC.

Cost estimation and screening of external MSAs   To determine which external MSA should be used to remove this load, it is necessary to determine the supply and target compositions as well as unit cost data for each MSA. Towards this end, one ought to consider the various processes undergone by each MSA. For instance, activated carbon, S_3, has an equilibrium relation (adsorption isotherm) for adsorbing phenol that is linear up to a lean-phase mass fraction of 0.11, after which activated carbon is quickly saturated and the adsorption isotherm levels off. Hence, x_3^t is taken as 0.11. It is also necessary to check the thermodynamic feasibility of this composition. Equation (3.5a) can be used to calculate the corresponding composition on the y scales,

$$y = 0.02(0.11 + 0.001) = 0.0022. \tag{3.13}$$

This value is less than the supply compositions of R_1 and R_2. Hence, it is possible to transfer phenol from both waste streams to S_3. Furthermore, this value is less than the composition at the tail of the lean composite stream (0.01). Therefore, S_3 will not eliminate any phenol that can be removed by the process MSAs (S_1 and S_2). After adsorbing phenol, activated carbon is fed to a steam regeneration column. In this unit, a mass ratio 2 : 1 of steam to adsorbed phenol is used to remove almost all adsorbed phenol. The regenerated activated carbon is recycled back to the MEN. During regeneration, 5% of the activated carbon is lost and has to be replenished by fresh makeup activated carbon. The steam leaving the regeneration column is condensed. The phenol in the condensate is separated via decantation followed by distillation. The cost of decantation and distillation is almost equal to the value of the recovered phenol. Hence, the operating cost of activated carbon (C_3, \$/kg of recirculating activated carbon) can be obtained as follows:

C_3 = cost of makeup + cost of regeneration

    = 0.05 × unit cost of fresh activated carbon \$/kg + 2 × cost of steam \$/kg

      × amount of phenol adsorbed (kg phenol/kg activated carbon)

Considering the unit costs of activated carbon and steam to be 1.60 and 5×10^{-3} \$/kg, respectively, we get

$$C_3 = 0.05 \times 1.60 + 2 \times 5 \times 10^{-3} x_3^t. \tag{3.14}$$

But, the mass of activated carbon needed to remove 1 kg of phenol from the waste streams can be evaluated through material balance,

$$1 \text{ kg of phenol} = \text{kg activated carbon} \left(x_3^t - 0 \right), \tag{3.15a}$$

i.e.,

    kg of activated carbon needed to remove 1 kg of phenol from waste streams

$$= 1/x_3^t. \tag{3.15b}$$

Hence, one can multiply Eqs. (3.14) and (3.15b) to obtain the cost of removing 1 kg of phenol from waste streams using activated carbon, C_3^r (\$/kg phenol removed) as

$$C_3^r = \frac{0.08}{x_3^t} + 0.01. \tag{3.16}$$

Substituting the value of x_3^t (0.11 kg phenol/kg activated carbon) into Eqs. (3.14) and (3.16), one obtains

$$C_3 = \$0.081/\text{kg recirculating activated carbon} \tag{3.17}$$

and

$$C_3^r = \$0.737/\text{kg of removed phenol} \qquad (3.18)$$

If the ion-exchange resin is used for removing phenol, it is regenerated by employing caustic soda to convert phenol into sodium phenoxide (a salable compound) according to the following reaction:

$$C_6H_5OH + NaOH \rightarrow C_6H_5ONa + H_2O$$

The reaction is thermodynamically favorable and proceeds to convert almost all the phenol to sodium phenoxide. Next, sodium phenoxide is separated from the caustic solution via distillation. The distillation cost is almost equal to the value of generated sodium phenoxide. Hence, the main operating costs of the process are the amount of sodium hydroxide that reacts with phenol and the cost of makeup resin. The molecular weights of phenol and sodium hydroxide are 94 and 40, respectively, so one can readily see that the amount of reacted sodium hydroxide is 0.426 (kg NaOH)/(kg phenol). The rate of ion-exchange resin deactivation is taken as 5%. Hence, the flowrate of resin makeup is 0.05 that of recirculating resin. Thus, the operating cost of the resin (C_4, \$/kg of recirculating resin) is determined through

$$
\begin{aligned}
C_4 &= \text{cost of makeup} + \text{cost of regeneration} \\
&= 0.05 \times \text{unit cost of fresh resin \$/kg} \\
&\quad + 0.426 \times \text{cost of sodium hydroxide \$/kg} \\
&\quad \times \text{amount of phenol adsorbed (kg phenol/kg resin)}
\end{aligned}
$$

Considering the unit costs of resin and sodium hydroxide to be 3.80 and 0.30 \$/kg, respectively, one obtains

$$
\begin{aligned}
C_4 &= 0.05 \times 3.80 + 0.426 \times 0.30x_4^t \\
&= 0.19 + 0.128x_4^t. \qquad (3.19)
\end{aligned}
$$

The mass of the resin needed to remove one kg of phenol from the waste streams can be determined from a material balance,

$$1 \text{ kg of phenol removed from waste streams} = \text{mass of resin} \left(x_4^t - 0\right). \quad (3.20)$$

Therefore, the cost of removing 1 kg of phenol from waste streams using ion-exchange resin, C_4^r(\$/kg phenol removed) is determined by combining Eqs. (3.19) and (3.20) to yield

$$C_4^r = \frac{0.19}{x_4^t} + 0.128. \qquad (3.21)$$

As Eq. (3.21) indicates, the higher the value of x_4^t, the lower the removal cost. So, what is the highest possible value of x_4^t that should be used? Since no mass is transferred across the pinch, material balance above the pinch is completely satisfied by using the process MSAs. By extending the target composition of S_4 to be above the pinch, one transfers a load that can be removed by process MSAs (for free) to S_4. Hence, when the operating cost of the network is considered, there is no economic incentive to pass S_4 above the pinch point. Therefore, the optimum target composition of S_4 is that corresponding to the pinch. Since the pinch composition on the waste scale is 0.0168, one can readily evaluate the corresponding composition for S_4 by applying Eq. (3.5b),

$$x_4^t = \frac{0.0168}{0.09} - 0.001 = 0.186. \tag{3.22}$$

By using this value, we can now invoke Eqs. (3.19) and (3.21) to get

$$C_4 = \$0.214/\text{kg of recirculating resin} \tag{3.23}$$

and

$$C_4^r = \$1.150/\text{kg of removed phenol} \tag{3.24}$$

When air stripping is used to remove phenol from a waste stream, the gaseous stream leaving the MEN cannot be discharged to the atmosphere owing to air-quality regulations. Hence, the air leaving the MEN is fed to a phenol-recovery unit in which a refrigerant is used to condense phenol. Based on the cooling duty, the cost of handling air in the condensation unit is

$$C_5 = \$0.06/\text{kg air} \tag{3.25}$$

The composition of phenol in the air leaving the MEN should be below the lower flammability limit. But, the LFL for phenol in air is 5.8 w/w%. An operating composition less than 50% of the LFL is typically suggested. Hence,

$$x_5^t = 0.5 \times 0.058 = 0.029 \tag{3.26}$$

This value corresponds to a corresponding y-scale composition of 0.0012, which is less than the supply mass fraction of phenol in either waste stream as well as the pinch composition. Hence, it is thermodynamically feasible for S_5 to recover phenol from R_1 and R_2. In addition, any amount of phenol removed by S_5 does not overlap with the load handled by the process MSAs. Therefore, the operating cost of air needed to remove 1 kg of phenol can be evaluated as follows:

$$C_5^r = 0.06/x_5^t = \$2.069/\text{kg of removed phenol} \tag{3.27}$$

Since it is thermodynamically feasible for any of the three external MSAs to remove the remaining phenolic load (0.0124 kg phenol/s), one should select the MSA that minimizes the operating cost. This is achieved by comparing the values of C_j^r reported by Eqs. (3.18), (3.24), and (3.27) to be 0.737, 1.150, and 2.069 \$/kg of removed phenol by using activated carbon, ion-exchange resin, and air, respectively. Therefore, activated carbon is the optimum external MSA. In addition, it is possible to evaluate the flowrate and operating cost of activated carbon as

$$L_3 = 0.0124/(0.11 - 0) = 0.1127 \text{ kg/s} \tag{3.28}$$

Also,

$$
\begin{aligned}
\text{Annual operating cost} \\
= \left(\frac{\$0.081}{kg \text{ activated carbon}} \right) \left(0.1127 \frac{kg \text{ activated carbon}}{s} \right) \\
\times \left(\frac{3600 \text{ s}}{hr} \right) \left(\frac{8760 \text{ hr}}{yr} \right) \\
= \$288 \times 10^3/yr.
\end{aligned}
\tag{3.29a}
$$

Equivalently, the same result can be obtained through the following equation:

$$
\begin{aligned}
\text{Annual operating cost} \\
= \left(\frac{\$0.737}{kg \text{ phenol removed}} \right) (0.0124) \left(\frac{kg \text{ phenol removed}}{s} \right) \\
\times \left(\frac{3600 \text{ s}}{hr} \right) \left(\frac{8760 \text{ hr}}{yr} \right) \\
= \$288 \times 10^3/yr.
\end{aligned}
\tag{3.29b}
$$

Once again, this MOC solution is a design target that has been identified even before determining the network configuration and equipment design. If this target is acceptable to the company, more detailed design efforts are warranted to scrutinize equilibrium as well as cost data, configure the system, and develop the detailed internal design of the units. This approach of "breadth first, depth later" is a key element in the overall design philosophy presented in this book.

3.7 Constructing Pinch Diagrams without Process MSAs

So far, an MOC solution has been identified through a two-stage process. First, the use of process MSAs is maximized by constructing the pinch diagram with the lean composite stream composed of process MSAs only. In the second stage, the external MSAs are screened to remove the remaining load at minimum cost.

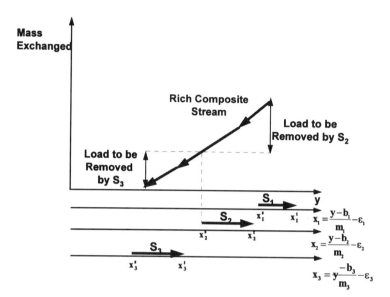

Figure 3.18a Screening external MSAs.

Suppose that the process does not have any process MSAs. How can a lean composite line be developed? The following shortcut method can be employed to construct the pinch diagram for external MSAs. A more rigorous method is presented in Chapter Six.

First, the rich composite line is plotted. Then, Eq. (3.5) is employed to generate the correspondence among the rich composition scale, y, and the lean composition scales for all external MSAs. Each external MSA is then represented versus its composition scale as a horizontal arrow extending between its supply and target compositions (Fig. 3.18a). Several useful insights can be gained from this diagram. Let us consider three MSAs; S_1, S_2 and S_3 whose costs ($/kg of recirculating MSA) are c_1, c_2 and c_3, respectively. These costs can be converted into $/kg of removed pollutant, c_j^r, as follows:

$$c_j^r = \frac{c_j}{x_j^t - x_j^s} \quad \text{where } j = 1, 2, 3. \quad (3.30)$$

If arrow S_2 lies completely to the left of arrow S_1 and c_2^r is less than c_1^r, one can eliminate S_1 from the problem since it is thermodynamically and economically inferior to S_2. On the other hand, if arrow S_3 lies completely to the left of arrow S_2 but c_3^r is greater than c_2^r, one should retain both MSAs. In order to minimize the operating cost of the network, separation should be staged to use the cheapest MSA

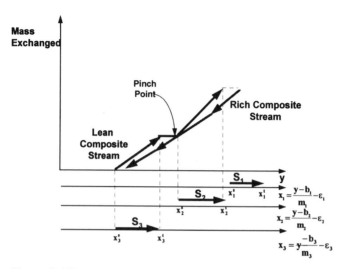

Figure 3.18b Constructing the pinch diagram for external MSAs.

where it is feasible. Hence, S_2 should be used to remove all the rich load to its left while the remaining rich load is removed by S_3 (Fig. 3.18a). The flowrates of S_2 and S_3 are calculated by simply dividing the richload removed by the composition difference for the MSA. Now that the MSAs have been screened and their optimal flowrates have been determined, one can construct the pinch diagram as shown in Fig. 3.18b.

Example 3.3: Toluene Removal from Wastewater

Toluene is to be removed from a wastewater stream. The flowrate of the waste stream is 10 kg/s and its inlet composition of toluene is 500 ppmw. It is desired to reduced the toluene composition in water to 20 ppmw. Three external MSAs are considered: air (S_2) for stripping, activated carbon (S_2) for adsorption, and a solvent extractant (S_3). The data for the candidate MSAs are given in Table 3.6. The equilibrium data for the transfer of the pollutant from the waste stream to the jth MSA is given by

$$y_l = m_j x_j \tag{3.31}$$

where y_l and x_j are the mass fractions of the toluene in the wastewater and the jth MSA, respectively.

Use the pinch diagram to determine the minimum operating cost of the MEN.

Table 3.6
Data for MSAs of Toluene Removal Example

Stream	Upper bound on flowrate L_j^C kg/s	Supply composition (ppmw) x_j^s	Target composition (ppmw) x_j^t	m_j	ε_j ppmw	C_j $/kg MSA
S_1	∞	0	19,000	0.0084	6,000	7.2×10^{-3}
S_2	∞	100	20,000	0.0012	15,000	0.11
S_3	∞	50	2,100	0.0040	10,000	0.09

Solution

Based on the given data and Eq. (3.30), we can calculate c_j^r to be 0.38, 5.53, and 43.90 $/kg of toluene removed for air, activated carbon, and the extractant, respectively. Since air is the least expensive, it will be used to remove all the load to its right (0.0045 kg toluene/s, as can be seen from Fig. 3.19a). Therefore, the flowrate of air can be calculated as

$$\text{Flowrate of air} = \frac{0.0045}{19,000 \times 10^{-6}} = 0.237 \text{ kg/s} \qquad (3.32)$$

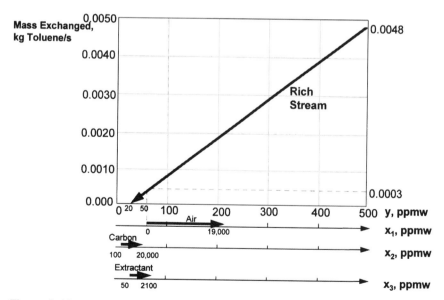

Figure 3.19a Screening external MSAs for the toluene-removal example.

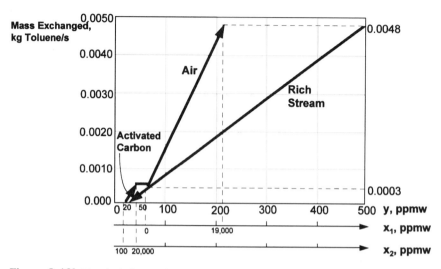

Figure 3.19b The pinch diagram for the toluene-removal example.

Since activated carbon lies to the left of the extractant and has a lower c_j^r, it will be used to remove the remaining load (0.0003 kg toluene/s, as shown by Fig. 3.19b). The flowrate of activated carbon is 0.015 kg/s. For 8000 operating hours per year, the annual operating cost of the system is $96,700/year. The annualized fixed cost can be calculated after equipment sizing (as shown in Example 2.1).

3.8 Trading Off Fixed versus Operating Costs

As has been mentioned in Section 3.4, two design targets can be determined for an MEN : MOC and minimum number of units (aimed at minimizing fixed cost). While this chapter has presented the pinch analysis as a method of determining the MOC target, the next two chapters provide systematic procedures for minimizing the fixed cost while realizing the MOC solution. Since the two targets are not necessarily compatible, the designer should trade off both costs. Two principal approaches may be used: varying driving forces and mixing waste streams.

As discussed in Section 2.5, the minimum allowable composition differences can be used to trade off fixed versus operating costs. Typically, an increase in ε_j leads to an increase in the MOC of the network (see Figs. 3.12 and 3.14) and a decrease in the fixed cost of the system. Hence, the minimum allowable composition differences can be iteratively varied until the total annualized cost of the system is attained (see Fig. 2.13).

The other common approach for trading off fixed costs against operating costs is mixing of waste streams. In some cases, the plant operation and the

Table 3.7
Data of Waste Streams for Dephenolization Example with Mixing of Waste Streams

Stream	Description	Flowrate G_i, kg/s	Supply composition (mass fraction) y_i^s	Target composition (mass fraction) y_i^t
R_{mixed}	Mixed R_1 and R_2	3	0.0433	0.0087

environmental regulations allow such mixing, typically decreases the number of mass exchangers and, consequently, the fixed cost. On the other hand, mixing various waste streams normally increases the MOC of the system. This can be explained on a pinch diagram because mixing results in a right shift of the waste composite stream. Such a shift may make a process MSA or a low-cost external MSA infeasible; increasing the operating cost of the MSAs. There are also cases when mixing waste streams does not affect the MOC solution. For instance, consider the previous dephenolization example. Suppose that process considerations allow mixing of the two waste streams. Furthermore, assume that the waste recovery task is described in terms of retaining an 80% phenol recovery from the mixed stream. The data for the mixed waste stream are given in Table 3.7. A pinch analysis (Fig. 3.20) demonstrates that the MOC solution is still the same. In

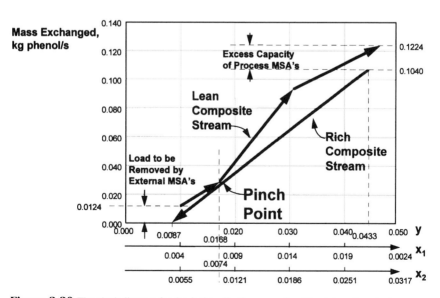

Figure 3.20 The pinch diagram for the dephenolization example with mixing of waste streams.

Table 3.8
Data for the Wastewater Stream of Tire Pyrolysis Plant

Stream	Description	Flowrate G_i, kg/s	Supply composition (ppmw) y_i^s	Target composition (ppmw) y_i^t
R_1	Aqueous layer from decanter	0.2	500	50

this case, mixing the waste streams is recommended since it reduces the number of mass exchangers without affecting the MOC solution.

Problems

3.1 A processing facility converts scrap tires into fuel via pyrolysis. Figure 3.21 is a simplified block flow diagram of the process. The discarded tires are fed to a high-temperature reactor where heat breaks down the hydrocarbon content of the tires into oils and gaseous fuels. The oils are further processed and separated to yield transportation fuels. The reactor off-gases are cooled to condense light oils. The condensate is decanted into two layers: organic and aqueous. The organic layer is mixed with the liquid products of the reactor. The aqueous layer is a wastewater stream whose organic content must be reduced prior to discharge. The primary pollutant in the wastewater is a heavy hydrocarbon. The data for the wastewater stream are given in Table 3.8. A process lean stream and three external MSAs are considered for removing the pollutant. The process lean stream is a flare gas (a gaseous stream fed to the flare) which can be used as a process stripping agent. To

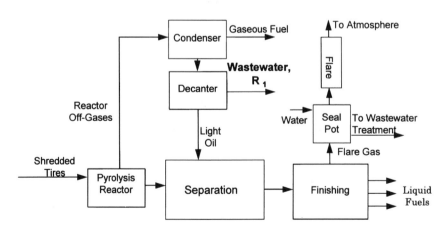

Figure 3.21 A simplified block flow diagram of a tire-to-fuel process.

Table 3.9
Data for the MSAs of the Tire Pyrolysis Problem

Stream	Upper bound on flowrate L_j^C kg/s	Supply composition (ppmw) x_j^s	Target composition (ppmw) x_j^t	m_j	ε_j ppmw	C_j $/kg MSA
S_1	0.15	200	900	0.5	200	–
S_2	∞	300	1000	1.0	100	0.001
S_3	∞	10	200	0.8	50	0.020
S_4	∞	20	600	0.2	50	0.040

prevent the back-propagation of fire from the flare, a seal pot is used. An aqueous stream is passed through the seal pot to form a buffer zone between the fire and the source of the flare gas. Therefore, the seal pot can be used as a stripping column in which the flare gas strips the organic pollutant off the wastewater while the wastewater stream constitutes a buffer solution for preventing back-propagation of fire.

Three external MSAs are considered: a solvent extractant (S_2), an adsorbent (S_3), and a stripping agent (S_4). The data for the candidate MSAs are given in Table 3.9. The equilibrium data for the transfer of the pollutant from the waste stream to the jth MSA is given by

$$y_1 = m_j x_j \tag{3.33}$$

where y_1 and x_j are the mass fractions of the organic pollutant in the wastewater and the jth MSA, respectively.

For the given data, use the pinch diagram to determine the minimum operating cost of the MEN.

3.2 If the fixed cost is disregarded in the previous problem, what is the lowest target for operating cost of the MEN? **Hint:** Set all the ε_j's equal to zero.

3.3 Consider the coke-oven gas COG sweetening process shown in Fig. 3.22. The basic objective of COG sweetening is the removal of acidic impurities, primarily hydrogen sulfide, from COG (a mixture of H_2, CH_4, CO, N_2, NH_3, CO_2, and H_2S). Hydrogen sulfide is an undesirable impurity, because it is corrosive and contributes to SO_2 emission when the COG is burnt. The existence of ammonia in COG and the selectivity of aqueous ammonia in absorbing H_2S suggests that aqueous ammonia is a candidate solvent (process lean stream, S_1). It is desirable that the ammonia recovered from the sour gas compensate for a large portion of the ammonia losses throughout the system and, thus, reduce the need for ammonia makeup. Besides ammonia, an external MSA (chilled methanol, S_2) is also available for service to supplement the aqueous ammonia solution as needed.

The purification of the COG involves washing the sour COG, R_1, with sufficient aqueous ammonia and/or chilled methanol to absorb the required amounts of hydrogen sulfide. The acid gases are subsequently stripped from the solvents and the regenerated

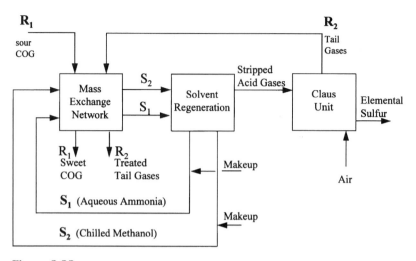

Figure 3.22 Sweetening of COG (from El-Halwagi and Manousiouthakis, 1989. Reproduced with permission of the American Institute of Chemical Engineers. Copyright © 1989 AIChE. All rights reserved).

MSAs are recirculated. The stripped acid gases are fed to a "Claus unit" where elemental sulfur is recovered from hydrogen sulfide. In view of air pollution control regulations, the tail gases leaving the Claus unit, R_2, should be treated for partial removal of the unconverted hydrogen sulfide. Table 3.10 summarizes the stream data.

Using the pinch diagram with $\varepsilon_1 = \varepsilon_2 = 0.0001$, find the minimum cost of MSAs required to handle the desulfurization of R_1 and R_2. Where is the pinch located?

3.4 Figure 3.23 is a simplified flow diagram of an oil refinery. The process generates two major sources of phenolic wastewater; one from the catalytic cracking unit and the other from the visbreaking system. Two technologies can be used to remove phenol from R_1 and R_2: solvent extraction using light gas oil S_1 (a process MSA) and adsorption using activated carbon S_2 (an external MSA). Table 3.11 provides data for the streams. A minimum allowable composition difference, ε_j, of 0.01 can be used for the two MSAs.

Table 3.10
Stream Data for the COG-Sweetening Problem

	Rich stream				MSAs					
Stream	G_i (kg/s)	y_i^s	y_i^t	Stream	L_j^c kg/s	x_j^s	x_j^t	m_j	b_j	c_j $/kg
R_1	0.90	0.0700	0.0003	S_1	2.3	0.0006	0.0310	1.45	0.000	0.00
R_2	0.10	0.0510	0.0001	S_2	∞	0.0002	0.0035	0.26	0.000	0.10

Table 3.11
Stream Data for Refinery Problem

	Rich stream				MSAs				
Stream	G_i (kg/s)	y_i^s	y_i^t	Stream	L_j^c kg/s	x_j^s	x_j^t	m_j	b_j	c_j $/kg
R_1	8.00	0.10	0.01	S_1	10.00	0.01	0.02	2.00	0.00	0.00
R_2	6.00	0.08	0.01	S_2	∞	0.00	0.11	0.02	0.00	0.08

By constructing a pinch diagram for the problem, find the minimum cost of MSAs needed to remove phenol from R_1 and R_2. How do you characterize the point at which both composite streams touch? Is it a true pinch point?

3.5 Figure 3.24 shows the process flowsheet for an ethylene/ethylbenzene plant. Gas oil is cracked with steam in a pyrolysis furnace to form ethylene, low BTU gases, hexane, heptane, and heavier hydrocarbons. The ethylene is then reacted with benzene to form ethylbenzene (Stanley and El-Halwagi, 1995). Two wastewater streams are formed: R_1

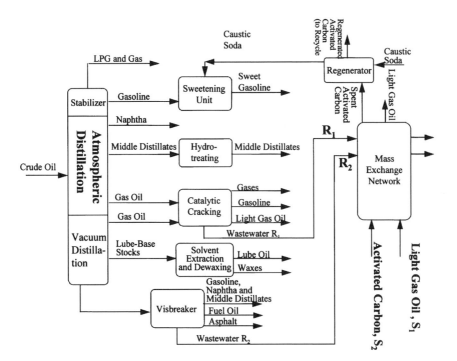

Figure 3.23 Dephenolization of refinery wastes (from El-Halwagi et al., 1992).

Table 3.12
Data for the Waste Streams of the Ethylbenzene Plant

Stream	Description	Flowrate G_i, kg/s	Supply composition (ppmw) y_i^s	Target composition (ppmw) y_i^t
R_1	Wastewater from settling	100	1,000	200
R_2	Wastewater from ethylbenzene separation	50	1,800	360

which is the quench water recycle for the cooling tower and R_2 which is the wastewater from the ethylbenzene portion of the plant. The primary pollutant present in the two wastewater streams is benzene. Benzene must be removed from Stream R_1 down to a concentration of 200 ppm before R_1 can be recycled back to the cooling tower. Benzene must also be removed from stream R_2 down to a concentration of 360 ppm before R_2 can be sent to biotreatment. The data for streams R_1 and R_2 are shown in Table 3.12.

There are two process MSAs available to remove benzene from the wastewater streams. These process MSAs are hexane (S_1) and heptane (S_2). Hexane is available at a flowrate of 0.8 kg/s and supply composition of 10 ppmw while heptane is available at a flowrate

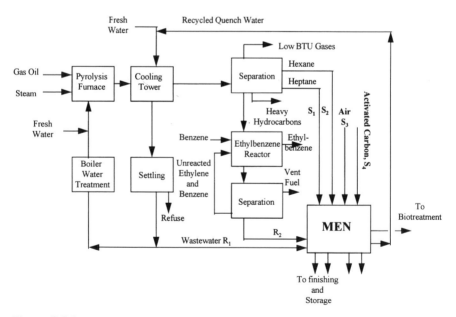

Figure 3.24 Process flowsheet for an ethylene/ethylbenzene plant.

of 0.3 kg/s and supply composition of 15 ppmw. The target compositions for hexane and heptane are unknown and should be determined by the engineer designing the MEN. The mass-transfer driving forces, ε_1 and ε_2, should be at least 30,000 and 20,000 ppmw, respectively. The equilibrium data for benzene transfer from wastewater to hexane and heptane are:

$$y = 0.011x_1 \qquad (3.34)$$

and

$$y = 0.008x_2, \qquad (3.35)$$

where y, x_1 and x_2 are given in mass fractions.

Two external MSAs are considered for removing benzene; air (S_3) and activated carbon (S_4). Air is compressed to 3 atm before stripping. Following stripping, benzene is separated from air using condensation. Henry's law can be used to predict equilibrium for the stripping process. Activated carbon is continuously regenerated using steam in the ratio of 1.5 kg steam : 1 kg of benzene adsorbed on activated carbon. Makeup at the rate of 1% of recirculating activated carbon is needed to compensate for losses due to regeneration and deactivation. Over the operating range, the equilibrium relation for the transfer of benzene from wastewater onto activated carbon can be described by:

$$y = 7.0 \times 10^{-4}x_4. \qquad (3.36)$$

(a) Using the pinch diagram determine the pinch location, minimum load of benzene to be removed by external MSAs and excess capacity of process MSAs. How do you remove this excess capacity?

(b) Considering the four candidate MSAs, what is the MOC needed to remove benzene?

3.6 Consider the magnetic-tape manufacturing process (Dunn et al., 1995) shown in Fig. 3.25. First, coating ingredients are dissolved in 0.09 kg/s of organic solvent and mixed to form a slurry. The slurry is suspended with resin binders and special additives. Next, the coating slurry is deposited on a base film. Nitrogen gas is used to induce evaporation

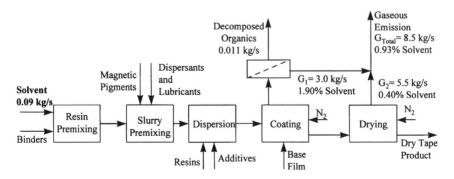

Figure 3.25 Schematic representation of a magnetic-tape manufacturing process.

Table 3.13
Data for the MSAs

Stream	Upper bound on flowrate L_j^C kg/s	Supply composition (mass fraction) x_j^s	Target composition (mass fraction) x_j^t	m_j	ε_j mass fraction	C_j $/kg MSA
S_1	∞	0.014	0.040	0.4	0.001	0.002
S_2	∞	0.020	0.080	1.5	0.001	0.001
S_3	∞	0.001	0.010	0.1	0.001	0.002

rate of solvent that is proper for deposition. In the coating chamber, 0.011 kg/s of solvent are decomposed into other organic species. The decomposed organics are separated from the exhaust gas in a membrane unit. The retentate stream leaving the membrane unit has a flowrate of 3.0 kg/s and is primarily composed of nitrogen that is laden with 1.9 wt/wt% of the organic solvent. The coated film is passed to a dryer, where nitrogen gas is employed to evaporate the remaining solvent. The exhaust gas leaving the dryer, has a flowrate of 5.5 kg/s and contains 0.4 wt/wt% solvent. The two exhaust gases are mixed and disposed off.

Due to environmental regulations, a reduction in the total solvent emission is required that is equal to 0.06 kg/s (i.e., removing 25% of current emission). Three MSAs can be used to remove the solvent from the gaseous emission. The equilibrium data for the transfer of the organic solvent to the jth lean stream is given by $y = m_j x_j$ where the values of m_j are given in Table 3.13. Throughout this problem, a minimum allowable composition difference, ε_j, of 0.001(kg organic solvent)/(kg MSA) is to be used. The data for the MSAs are given in Table 3.13.

(a) Using the pinch diagram, determine which solvent(s) should be employed to remove the solvent? What is the MOC for the solvent removal task? **Hint:** Consider segregating the two waste streams and removing solvent from one of them. The annualized fixed cost of a mass exchanger, $/yr, may be approximated by 18,000 (Gas Flowrate, kg/s)$^{0.65}$.

(b) The value of the recovered solvent is $0.80/kg of organic solvent. What is the annual gross revenue (annual value of recovered solvent − total annualized cost of solvent recovery system)?

Symbols

b_j intercept of equilibrium line for the jth MSA

C_j unit cost of the jth MSA including regeneration and makeup, $/unit flowrate of recirculating MSA

C_j^r unit cost of the jth MSA required to remove a unit mass/mole of the key pollutant,

G_i Flowrate of the ith waste stream

i index for waste streams

j	index for MSAs
L_j	flowrate of the jth MSA
L_j^C	upper bound on the available flowrate of the jth MSA
m_j	slope of equilibrium line for the jth MSA
MR_i	mass/moles of pollutant lost from the ith waste stream as defined by Eq. (3.6)
MS_j	mass/moles of pollutant gained by the jth MSA as defined by Eq. (3.7)
N_i	number of independent synthesis problems
N_S	number of MSAs
N_{SE}	number of external MSAs
N_{SP}	number of process MSAs
N_R	number of rich (waste) streams
R_i	the ith waste stream
S_j	the jth MSA
U	minimum number of mass-exchange units, defined by Eq. (3.2)
x_j	composition of key component in the jth MSA
x_j^{in}	inlet composition of key component in the jth MSA
$x_j^{in,\,max}$	maximum practically feasible inlet composition of key component in the jth MSA
$x_j^{in,\,*}$	maximum thermodynamically feasible inlet composition of key component in the jth MSA
x_j^{out}	outlet composition of key component in the jth MSA
x_j^s	supply composition of the key component in the jth MSA
x_j^t	target composition of the key component in the jth MSA
x_j^*	composition of key component in the jth MSA which is in equilibrium with y_i
y	composition scale for the key component in any waste stream
y_i	composition of key component in the ith waste stream
y_i^s	supply composition of key component in the ith waste stream
y_i^t	target composition of key component in the ith waste stream

Greek

ε_j	minimum allowable composition difference for the jth MSA

References

Allen, D. T., Bakshani, N., and Rosselot, K. S. (1992). "Pollution Prevention—Homework and Design Problems for Engineering Curricula," *Am. Inst. Chem. Eng.*, New York.

Dhole, V. R., Ramchandani, N., Tainsh, R. A., and Wasilewski, M. (1996). Make your process water pay for itself. *Chem. Eng.*, January, pp. 100–103.

Dunn, R. F., and El-Halwagi, M. M. (1993). Optimal recycle/reuse policies for minimizing the wastes of pulp and paper plants. *Environ. Sci. Health* **A28**(1), 217–234.

Dunn, R. F., El-Halwagi, M. M., Lakin, J., and Serageldin, M. (1995). Selection of Organic Solvent Blends for Environmental Compliance in the Coating Industries. Proceedings of the First International Plant Operations and Design Conference, eds. E. D. Griffith, H. Kahn and M. C. Cousins, Vol. III, pp. 83–107, AIChE, New York.

El-Halwagi A. M., and El-Halwagi, M. M. (1992). Waste minimization via computer aided chemical process synthesis—A new design philosophy. *TESCE J.* **18**(2), 155–187.

El-Halwagi, M. M., and Manousiouthakis, V. (1989). Synthesis of mass exchange networks. *AIChE J.* **35**(8), 1233–1244.

El-Halwagi, M. M. (1993). A process synthesis approach to the dilemma of simultaneous heat recovery, waste reduction and cost effectiveness. In "Proceedings of the Third Cairo International Conference on Renewable Energy Sources" (A. I. El-Sharkawy and R. H. Kummler, eds.), Vol. 2, pp. 579–594 .

El-Halwagi, M. M., El-Halwagi, A. M., and Manousiouthakis, V. (1992). Optimal design of dephenolization networks for petroleum-refinery wastes. *Trans. Inst. Chem. Eng.* **70**, Part B,131–139.

El-Halwagi, M. M., and Manousiouthakis, V. (1990a). Automatic synthesis of mass exchange networks with single-component targets. *Chem. Eng. Sci.* **45**(9), 2813–2831.

El-Halwagi, M. M., and Manousiouthakis, V. (1990b). Simultaneous synthesis of mass exchange and regeneration networks. *AIChE J.* **36**(8), 1209–1219.

El-Halwagi, M. M., and Srinivas, B. K. (1992). Synthesis of reactive mass-exchange networks. *Chem. Eng. Sci.* **47**(8), 2113–2119.

Garrison, G. W., Cooley, B. L., and El-Halwagi, M. M. (1995). Synthesis of mass exchange networks with multiple target mass separating agents. *Dev. Chem. Eng. Miner. Proc.* **3**(1), 31–49.

Gupta, A., and Manousiouthakis, V. (1993). Minimum utility cost of mass exchanger networks with variable single component supplies and targets. *Ind. Eng. Chem. Res.* **32**(9), 1937–1950.

Huang, Y. L., and Edgar, T. F. (1995). Knowledge based design approach for the simultaneous minimization of waste generation and energy consumption in a petroleum refinery. In "Waste Minimization Through Process Design" (A. P. Rossiter, ed.), pp. 181–196. McGraw Hill, New York.

Huang, Y. L., and Fan, L. T. (1995). Intelligent process design and control for in-plant waste minimization. In "Waste Minimization Through Process Design" (A. P. Rossiter, eds.), pp.165–180. McGraw Hill, New York.

Kiperstok, A., and Sharratt, P. N. (1995). On the optimization of mass exchange networks for removal of pollutants. *Trans. Inst. Chem. Eng.* **73**, Part B, 271–277.

Papalexandri, K. P., and Pistikopoulos, E. N. (1994). A multiperiod MINLP model for the synthesis of heat and mass exchange networks. *Comput. Chem. Eng.* **18**(12), 1125–1139.

Samdani, G. S. (1995). Cleaner by design. *Chem. Eng.* July, pp. 32–37.

Srinivas, B. K., and El-Halwagi, M. M. (1994a). Synthesis of reactive mass-exchange networks with general nonlinear equilibrium functions. *AIChE J.* **40**(3), 463–472.

Srinivas, B. K., and El-Halwagi, M. M. (1994b). Synthesis of combined heat reactive mass-exchange networks. *Chem. Eng. Sci.* **49**(13), 2059–2074.

Stanley, C., and El-Halwagi, M. M. (1995). Synthesis of mass exchange networks using linear programming techniques. In "Waste Minimization Through Process Design" (A. P. Rossiter, ed.), pp. 209–224. McGraw Hill, New York.

Wang, Y. P., and Smith, R. (1994). Wastewater minimization. *Chem. Eng. Sci.* **49**(7), 981–1006.

Warren, A., Srinivas, B. K., and El-Halwagi, M. M. (1995). Design of cost-effective waste-reduction systems for synthetic fuel plants. *J. Environ. Eng.* **121**(10), 742–747.

Zhu, M., and El-Halwagi, M. M. (1995). Synthesis of flexible mass exchange networks. *Chem. Eng. Commun.* **138**, 193–211.

Graphical Techniques for Mass Integration with Mass-Exchange Interception

As has been discussed in Chapter One, interception is a key element of mass integration. Effective pollution prevention can be achieved by a combination of stream segregation, interception, recycle from sources to sinks (with or without interception) and sink/generator manipulation. Chapter Three has presented graphical tools for synthesizing MEN's which use MSA-induced interception. The MEN can be used to prepare sources for recycle. Before we address integrating recycle with interception, it is useful to consider direct recycle without interception. The current chapter illustrates how MEN synthesis can be incorporated within a more comprehensive mass-integration analysis. Since this chapter focuses on interception using MSA's, the waste interception network "WIN" of Fig. 1.4 becomes a MEN (Fig. 4.1). This is a representation of the flowsheet from a species viewpoint. It also provides a framework for identifying the solution strategies including segregation, mixing, interception, recycle and unit manipulation. Several graphical techniques will be introduced to demonstrate the integration of interception with other mass-integration strategies. For more rigorous techniques the reader is referred to Chapter Seven and literature (e.g., El-Halwagi et al., 1996; El-Halwagi and Spriggs, 1996; Garrison et al., 1995, 1996; Hamad et al., 1995, 1996).

4.1 The Source–Sink Mapping Diagram

As mentioned in Chapter One, recycle refers to the utilization of a pollutant-laden stream (a source) in a process unit (a sink). A source may be recycled to a sink

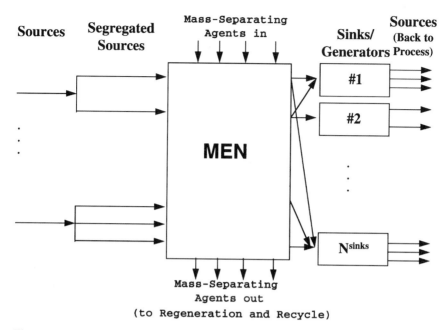

Figure 4.1 A schematic representation of mass-integration framework with MSA-induced interception (a WIN becomes a MEN) (from El-Halwagi et al., 1996. Reproduced with permission of the American Institute of Chemical Engineers. Copyright © 1996 AIChE. All rights reserved).

directly or following segregation and/or interception. Furthermore, multiple sources may be mixed prior to recycle. Before integrating interception with recycle, it is useful to discuss direct-recycle opportunities. For the same piping and pumping requirements, direct recycle is less expensive than recycle following interception. The source–sink mapping diagram is a visualization tool which can be used to determine direct-recycling opportunities (Fig. 4.2). For each pollutant, a diagram is constructed by plotting the pollutant load (flowrate × composition) or flowrate versus composition.

On the source–sink mapping diagram, sources are represented by shaded circles and sinks are represented by hollow circles. Typically, process constraints limit the range of pollutant composition and load that each sink can accept. The intersection of these two bands provides a zone of acceptable composition and load for recycle. If a source (e.g., source a) lies within this zone, it can be directly recycled to the sink (e.g., sink S). Moreover, sources b and c can be mixed using the lever-arm principle to create a mixed stream that can be recycled to sink S.

The source–sink mapping diagram can also be used to determine the extent of interception needed. If a source lies to the right of a sink, it can be intercepted to render it within the band of acceptable recycle. The problem of simultaneously

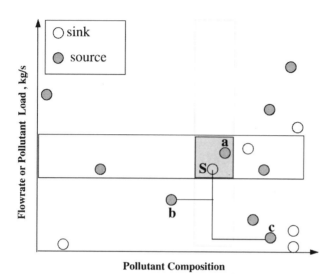

Figure 4.2 Identification of recycle opportunities using source–sink mapping diagram (from El-Halwagi and Spriggs, 1996).

intercepting several sources using MSA's is a MEN-synthesis task. The foregoing issues are illustrated in the next example.

4.2 Application of Mass Integration to Enhance Yield, Debottleneck the Process and Reduce Wastewater in an Acrylonitrile "an" Plant

Acrylonitrile (AN, C_3H_3N) is manufactured via the vapor-phase ammoxidation of propylene:

$$C_3H_6 + NH_3 + 1.5O_2 \xrightarrow{catalyst} C_3H_3N + 3H_2O.$$

The reaction takes place in a fluidized-bed reactor in which propylene, ammonia and oxygen are catalytically reacted at 450°C and 2 atm. The reaction is a single pass with almost complete conversion of propylene. The reaction products are cooled using an indirect-contact heat exchanger which condenses a fraction of the reactor off-gas. The remaining off-gas is scrubbed with water, then decanted into an aqueous layer and an organic layer. The organic layer is fractionated in a distillation column under slight vacuum which is induced by a steam-jet ejector. Figure 4.3 shows the process flowsheet along with pertinent material balance data.

The wastewater stream of the plant is composed of the off-gas condensate, aqueous layer of the decanter, bottom product of the distillation column and the

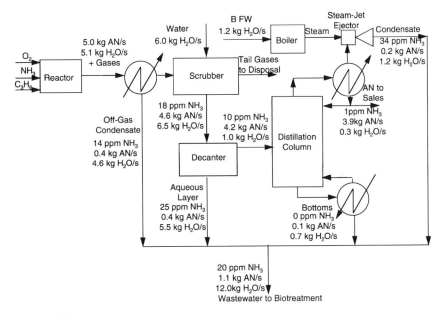

Figure 4.3 Flowsheet of AN production.

condensate from the steam-jet ejector. This wastewater stream is fed to the bio-treatment facility. Since the biotreatment facility is currently operating at full hydraulic capacity, it constitutes a bottleneck for the facility. Plans for expanding production of AN are contingent upon the debottlenecking of the biotreatment facility by reducing its influent or installing an additional treatment unit. The new biotreatment facility will cost about $4 million in capital investment and $360,000/year in annual operating cost, leading to a TAC of $760,000/year with a ten-year linear depreciation. The objective of this case study is to use mass integration techniques to devise cost-effective strategies to debottleneck the biotreatment facility.

The following technical constraints should be observed in any proposed solution:

Scrubber

- $5.8 \leq$ flowrate of wash feed (kg/s) ≤ 6.2.
- $0.0 \leq$ ammonia content of wash feed (ppm NH_3) ≤ 10.0.

Boiler feed water "BFW"

- Ammonia content of BFW (ppm NH_3) $= 0.0$.
- AN content of BFW (ppm AN) $= 0.0$.

Table 4.1
Data for the MSAs of the AN Problem

Stream	Upper bound on flowrate L_j^C	Supply composition (ppmw) x_j^s	Target composition (ppmw) x_j^t	m_j	ε_j ppmw	C_j \$/kg MSA	C_j^r \$/kg NH$_3$ removed
S$_1$	∞	0	6	1.4	2	0.004	667
S$_2$	∞	10	400	0.04	5	0.070	180
S$_3$	∞	3	1100	0.02	5	0.100	91

Decanter

· $10.6 \leq$ flowrate of feed (kg/s) ≤ 11.1.

Distillation column

· $5.2 \leq$ flowrate of feed (kg/s) ≤ 5.7.
· $0.0 \leq$ Ammonia content of feed (ppm NH$_3$) ≤ 30.0.
· $80.0 \leq$ AN content of feed (wt% AN) ≤ 100.0.

Furthermore, for quality and operability objectives the plant does not wish to recycle the AN product stream (top of distillation column), the feed to the distillation column and the feed to the decanter.

Three external MSA's are considered for removing ammonia from water; air (S$_1$), activated carbon (S$_2$) and an adsorbing resin (S$_3$). The data for the candidate MSA's are given in Table 4.1. The equilibrium data for the transfer of the pollutant from the waste stream to the jth MSA is given by,

$$y_l = m_j x_j, \tag{4.1}$$

where y_l and x_j are weight-based ppm's of ammonia in the wastewater and the jth MSA, respectively.

Solution

The first step in the analysis is to identify the target for debottlenecking the biotreatment facility. An overall water balance for the plant (Fig. 4.4) can be written as follows:

Water in + Water generated by chemical reaction

= Wastewater out + Water losses

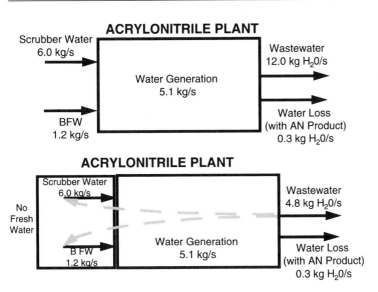

Figure 4.4 Establishing targets for biotreatment influent: overall water balance (a) before and (b) after mass integration.

Using mass-integration strategies of segregation, recycle, interception and sink/source manipulation, fresh-water usage can in principle be completely eliminated. Hence, for the same reaction conditions and water losses, the target for wastewater discharge can be calculated from the overall water balance as follows (Fig. 4.4):

> *Target of minimum discharge to biotreatment* $= 5.1 - 0.3$
>
> $$= 4.8 \text{ kg water/s.} \qquad (4.2)$$

Having identified this target, let us now determine the strategies needed to attain this target. In order to determine segregation and direct recycle opportunities, the sources and sinks should be examined. Once the streams composing the terminal wastewater are segregated, we get four sources that can be potentially recycled. Fresh water used in the scrubber and the boiler provide two more sources. In order to reduce wastewater discharge to biotreatment, fresh water must be reduced. Hence, we should focus our attention on recycling opportunities to sinks that employ fresh water; namely the scrubber and the boiler. Figure 4.5 illustrates the sources and sinks involved in the analysis.

Due to the stringent limitation on the BFW (no ammonia or AN), no recycled stream can be used in lieu of fresh water (segregation, recycle and interception can reduce but not eliminate ammonia/AN content). Hence, the boiler should not be considered as a sink for recycle (with or without interception). Instead, it should be handled at the stage of sink/generator manipulation. This leaves us with the five

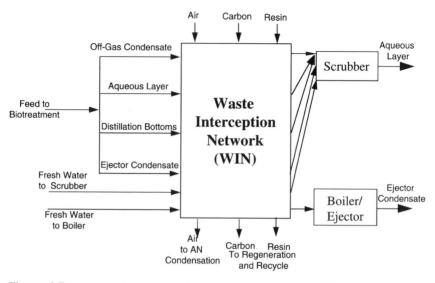

Figure 4.5 Segregation, interception and recycle representation for the AN case study.

segregated sources and one sink (scrubber) for the source-sink mapping diagram (Fig. 4.6).

In order to reduce fresh-water consumption in the scrubber, the usage of distillation bottoms and the off-gas condensate should be maximized since they have the least ammonia content. The flowrate resulting from combining these two sources (5.8 kg/s) is sufficient to run the scrubber. However, its ammonia composition[1] as determined by the lever-arm principle is 12 ppm, which lies outside the zone of permissible recycle to the scrubber. As shown by Fig. 4.7, the maximum flowrate of the off-gas condensate to be recycled to the scrubber[2] is determined to be 4.1 kg/s and the flowrate of fresh water is 0.9 kg/s (5.8 − 0.8 − 4.1). Therefore, direct recycle can reduce the fresh-water consumption (and consequently the

---

[1] Algebraically, this composition can be calculated as follows:

$$\frac{(5.0\,\text{kg/s})(14\,\text{ppm NH}_3) + 0}{5.8\,\text{kg/s}} = 12\,\text{ppm NH}_3$$

[2] Again, algebraically this flowrate can be calculated as follows:

$$\frac{\textit{Flowrate}\ \text{of recycled off} - \text{Gas condensate} \times 14\ \textit{ppm NH}_3 + 0 + 0}{5.8\ \textit{kg/s}} = 10\,\textit{pmm NH}_3.$$

where the numerator represents the ammonia in recycled off-gas condensate, distillation bottoms (none) and fresh water (none). Hence, flowrate of recycled off-gas condensate = 4.14 kg/s.

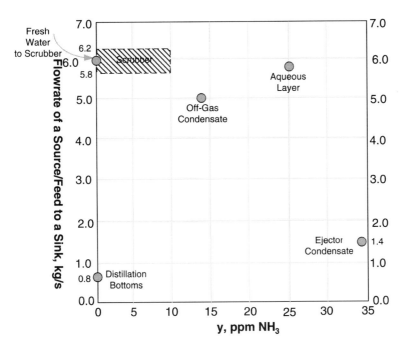

Figure 4.6 Source–sink mapping diagram for the AN case study.

influent to biotreatment) by 5.1 kg/s.

$$\frac{Arm\ of\ Gas\ Condensate}{Total\ Arm} = \frac{Flowrate\ of\ Recycled\ Gas\ Condensate}{Flowrate\ of\ Scrubber\ Feed}$$

i.e.,   (4.3)

$$\frac{10-0}{14-0} = \frac{Flowrate\ of\ Recycled\ Gas\ Condensate}{5.8}$$

The primary cost of direct recycling is pumping and piping. Assuming that the TAC for pumping and piping is $80/m · yr and assuming that the total length of piping is 600 m, the TAC for pumping and piping is $48,000/year.

Next, we include interception into the analysis. In order to eliminate fresh water from the scrubber, the composition of ammonia in the off-gas condensate must be reduced from 14 ppm to 12 ppm. This result may be obtained graphically as shown in Fig. 4.8. Alternatively, it may be calculated as follows:

$$\frac{(5.0\,kg/s)\,y^t\,ppm\ NH_3 + 0}{5.8\,kg/s} = 10\,ppm\ NH_3$$

i.e.,   (4.4)

$$y^t = 12\,ppm.$$

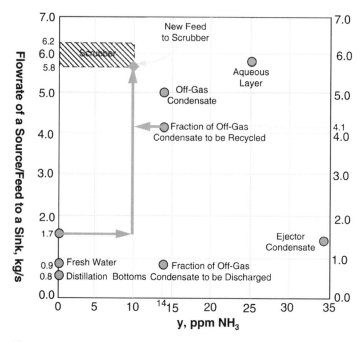

Figure 4.7 Direct-recycle opportunities for the AN case study.

In order to synthesize an optimal MEN for intercepting the off-gas condensate, we construct the pinch diagram as shown in Fig. 4.9. Since the three MSA's lie completely to the left of the rich stream, they are all thermodynamically feasible. Hence, we choose the one with the least cost ($/kg NH_3 removed); namely the resin. The annual operating cost for removing ammonia using the resin is:

$$5\frac{\text{kg Liquid}}{\text{s}} \times (14 \times 10^{-6} - 12 \times 10^{-6})\frac{\text{kg } NH_3}{\text{kg Liquid}} \times 91\frac{\$}{\text{kg } NH_3}$$

$$\times 3600 \times 8760\frac{\text{s}}{\text{yr}} = \$29,000/\text{yr}. \tag{4.5}$$

The annualized fixed cost of the adsorption column along with its ancillary equipment (e.g., regeneration, materials handling, etc.) is estimated to be about $90,000/yr. Therefore, the TAC for the interception system is $119,000/yr.

As a result of segregation, interception and recycle, we have eliminated the use of fresh water in the scrubber leading to a reduction of fresh water consumption (and influent to biotreatment) by 6.0 kg/s. Therefore, the target for segregation, interception and recycle has been realized. Next, we focus our attention on sink/generator manipulation to remove fresh-water consumption in the steam-jet

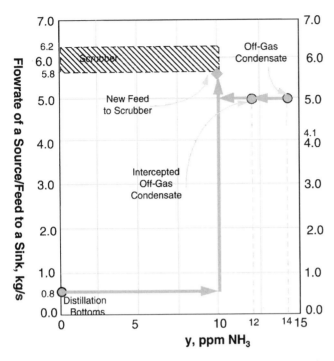

Figure 4.8 Interception and recycle opportunities for the AN case study.

ejector. The challenge here is to alter the design and/or operation of the boiler, the ejector or the distillation column to reduce or eliminate the use of steam. Several solutions may be proposed including:

• Replacement of the steam-jet ejector with a vacuum pump. The distillation operation will not be affected. The operating cost of the ejector and the vacuum pump are comparable. However, a capital investment of $75,000 is needed to purchase the pump. For a five-year linear depreciation with negligible salvage value, the annualized fixed cost of the pump is $15,000/year.

• Operating the column under atmospheric pressure thereby eliminating the need for the vacuum pump. Here a simulation study is needed to examine the effect of pressure change.

• Relaxing the requirement on BFW quality to few ppm's of ammonia and AN. In this case, recycle and interception techniques can be used to significantly reduce the fresh water feed to the boiler and, consequently, the net wastewater generated.

Figure 4.10 illustrates the revised flowsheet with segregation, interception, recycle and sink/generator manipulation. As can be seen from the figure, the flowrate of the terminal wastewater stream has been reduced to 4.8 kg H_2O/s. This is exactly the same *target* predicted in Fig. 4.4. In order to refine the

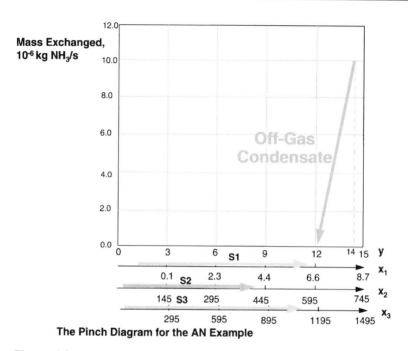

The Pinch Diagram for the AN Example

Figure 4.9 The pinch diagram for the AN case study.

material balance throughout the plant, a simulation study is needed, as discussed in Chapter 1.

Figures 4.11 and 4.12 are impact diagrams (sometime referred to as Pareto charts) for the reduction in wastewater and the associated TAC. These figures illustrate the cumulative impact of the identified strategies on biotreatment influent and cost.

We are now in a position to discuss the merits of the identified solutions. As can be inferred from Fig. 4.11, the following benefits can be achieved:

• Acrylonitrile production has increased from 3.9 kg/s to 4.6 kg/s, which corresponds to an 18% yield enhancement for the plant. This production increase is a result of better allocation of process streams; the essence of mass integration. For a selling value of $0.6/kg of AN, the additional production of 0.7 kg AN/s can provide an annual revenue of $13.3 million/yr!

• Fresh-water usage and influent to biotreatment facility are decreased by 7.2 kg/s. The value of fresh water and the avoidance of treatment cost are additional benefits.

• Influent to biotreatment is reduced to 40% of current level. Therefore, the plant production can be expanded 2.5 times the current capacity before the biotreatment facility is debottlenecked again.

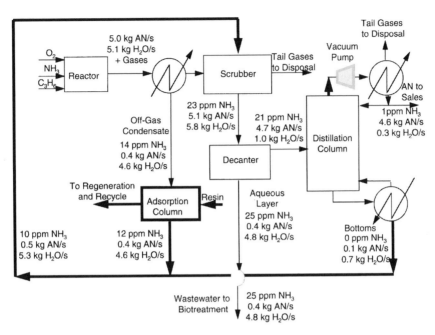

Figure 4.10 Optimal solution to the AN case study with segregation, recycle, interception and sink/generator manipulation.

Clearly, this is a superior solution to the installation of an additional biotreatment facility.

It is instructive to draw some conclusions from this case study and emphasize the design philosophy of mass integration. First, the target for debottlenecking the biotreatment facility was determined ahead of design. Then, systematic tools were used to generate optimal solutions that realize the target. Next, an analysis study is needed to refine the results. This is an efficient approach to understanding the global insights of the process, setting performance targets, realizing these targets and saving time and effort as a result of focusing on the big picture first and then dealing with the details later. This is a fundamentally different approach than using the designer's subjective decisions to alter the process and check the consequences using detailed analysis. It is also different from using simple end-of-pipe treatment solutions. Instead, the various species are optimally allocated throughout the process. Therefore, objectives such as yield enhancement, pollution prevention and cost savings can be simultaneously addressed. Indeed, pollution prevention (when undertaken with the proper techniques) can be a source of profit for the company, not an economic burden.

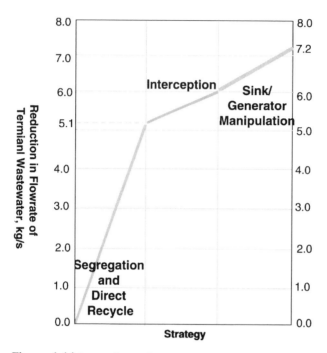

Figure 4.11 Impact diagram for reducing biotreatment influent.

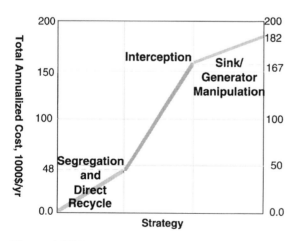

Figure 4.12 Impact diagram for TAC of identified solutions.

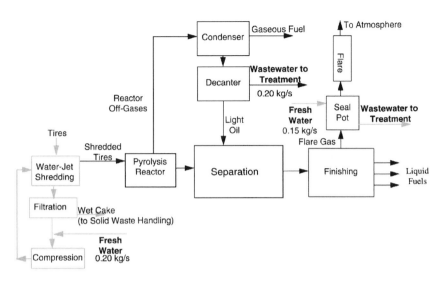

Figure 4.13 Schematic flow sheet of tire-to-fuel process.

Problems

4.1 Let us revisit the tire-to-fuel process described in Problem 3.1. Figure 4.13 is a more detailed flowsheet. Tire shredding is achieved by using high-pressure water jets. The shredded tires are fed to the process while the spent water is filtered. The wet cake collected from the filtration system is forwarded to solid waste handling. The filtrate is mixed with 0.20 kg/s of fresh water makeup to compensate for water losses with the wet cake (0.08 kg water/s) and the shredded tires (0.12 kg water/s). The mixture of filtrate and water makeup is fed to a high-pressure compression station for recycling the shredding unit. Due to the pyrolisis reactions, 0.08 kg water/s is generated.

The plant has two primary sources for wastewater, the decanter (0.20 kg water/s) and the seal pot (0.15 kg/s). The plant has been shipping the wastewater for off-site treatment. The cost of wastewater transportation and treatment is $0.01 kg leading to a wastewater treatment cost of approximately $110,000/yr. The plant wishes to stop off-site treatment of wastewater to avoid the cost ($110,000/yr) and alleviate legal-liability concerns in case of transportation accidents or inadequate treatment of the wastewater. The objective of this problem is to eliminate or reduce to the extent feasible off-site wastewater treatment. For capital budget authorization, the plant has the following economic criterion (see Appendix III):

$$\text{Payback period} = \frac{\textit{Fixed capital investment}}{\textit{Annual savings}} \le 3 \, \text{years}, \qquad (4.6)$$

where

Annual Savings = Annual avoided cost of off-site treatment
− Annual operating cost of on-site system

In addition to the information provided by Problem 3.1, the following data are available:

Economic Data

- Fixed cost of extraction system associated with S_2, $ = 120,000$ (Flowrate of wastewater, kg/s)$^{0.58}$
- Fixed cost of adsorption system associated with S_3, $ = 790,000$ (Flowrate of wastewater, kg/s)$^{0.70}$
- Fixed cost of stripping system associated with S_4, $ = 270,000$ (Flowrate of wastewater, kg/s)$^{0.65}$
- A biotreatment facility that can handle 0.35 kg/s wastewater has a fixed cost of $240,000 and an annual operating cost of $60,000/yr.

Technical Data

Water may be recycled to two sinks: the seal pot and the water-jet compression station. The following constraints on flowrate and composition of the pollutant (heavy organic) should be satisfied:

Seal plot

- $0.10 \leq$ Flowrate of feed water (kg/s) ≤ 0.20
- $0 \leq$ Pollutant content of feed water (ppmw) ≤ 500

Makeup to water-jet compression station

- $0.18 \leq$ Flowrate of makeup water (kg/s) ≤ 0.20
- $0 \leq$ Pollutant content of makeup water (ppmw) ≤ 50

4.2 Consider the magnetic-tape manufacturing process previously described by Problem 3.6. In this process coating ingredients are dissolved in 0.09 kg/s of organic solvent and mixed to form a slurry. The slurry is suspended with resin binders and special additives. Next, the coating slurry is deposited on a base film. Nitrogen gas is used to induce evaporation rate of solvent that is proper for deposition. In the coating chamber, 0.011 kg/s of solvent is decomposed into other organic species. The decomposed organics are separated from the exhaust gas in a membrane unit. The retentate stream leaving the membrane unit has a flowrate of 3.0 kg/s and is primarily composed of nitrogen that is laden with 1.9 wt/wt% of the organic solvent. The coated film is passed to a dryer where nitrogen gas is employed to evaporate the remaining solvent. The exhaust gas leaving the dryer has a flowrate of 5.5 kg/s and contains 0.4 wt/wt% solvent. The two exhaust gases are mixed and disposed off.

Three MSAs can be used to remove the solvent from the gaseous emission. The equilibrium data for the transfer of the organic solvent to the jth lean stream is given by $y = m_j x_j$, where the values of m_j are given in Table 4.2. Throughout this problem, a minimum allowable composition difference, ε_j, of 0.001 (kg organic solvent)/(kg MSA) is to be used. The data for the MSAs are given in Table 1.

The annualized fixed cost of a mass exchanger, $/yr, may be approximated by 18,000 (Gas Flowrate, kg/s)$^{0.65}$. The value of the recovered solvent is $0.80/kg of organic solvent.

Table 4.2
Data for the MSAs

Stream	Upper bound on flowrate L_j^C kg/s	Supply composition (mass fraction) x_j^s	Target composition (mass fraction) x_j^t	m_j	ε_j Mass fraction	C_j $/kg MSA
S_1	∞	0.014	0.040	0.4	0.001	0.002
S_2	∞	0.020	0.080	1.5	0.001	0.001
S_3	∞	0.001	0.010	0.1	0.001	0.002

In addition to the environment problem, the facility is concerned about the waste of resources, primarily in the form of used solvent (0.09 kg/s) that costs about $2.3 million/yr. It is desired to undertake a mass-integration analysis to optimize solvent usage, recovery and losses. It is possible to reuse solvent-laden gases in lieu of nitrogen gas in the coating and drying chambers. The following constraints on the gaseous feed to these two units should be observed:

Coating

- $3.0 \leq$ flowrate of gaseous feed (kg/s) ≤ 3.2.
- $0.0 \leq$ wt% of solvent ≤ 0.2.

Dryer

- $5.5 \leq$ flowrate of gaseous feed (kg/s) ≤ 6.0.
- $0.0 \leq$ wt% of solvent ≤ 0.1.

It may be assumed that outlet gas composition from the coating and the dry chambers are independent of the entering gas compositions.

(a) Determine the target for minimizing fresh solvent usage in the process.

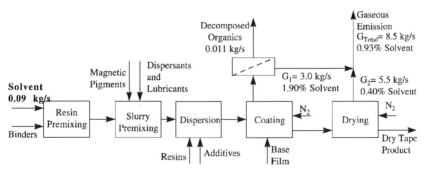

Figure 4.14 Schematic representation of a magnetic tape manufacturing process.

(b) Using segregation, mixing and direct recycle (without interception), what is the minimum consumption of nitrogen gas that should be used in the process? **Hint:** consider mixing fresh nitrogen with one or more of the exhaust gases prior to recycle.

(c) Develop a minimum-cost solution which minimizes the usage of fresh solvent in the process. The solution strategies may include segregation, mixing, recycling, and interception. Evaluate the paypack period for your solution.

4.3 Pulp and paper mills employ high levels of fresh water that lead to the generation of a significant amount of aqueous effluent. Therefore, the objective of optimizing water usage and wastewater discharge presents a major challenge to the industry. This objective calls for the use of mass integration techniques for the simultaneous source reduction, recycling, reuse, and treatment of aqueous wastes.

Due to the direct contact of water with various species, the aqueous streams are laden with various compounds including methanol, non-process elements "NPEs," and organic and inorganic species. In this problem, we focus on methanol as the primary species in water. Methanol is classified as a high priority pollutant for the pulping industry. In addition, it may provide a source of revenue if properly recovered.

The basic features of a typical continuous Kraft pulping process (Hamad et al., 1995) are shown in Fig. 4.15. Wood chips (containing 50% water) are conveyed from a surge hopper to a presteaming to facilitate subsequent impregnation of the chips with chemicals. A high-pressure feeder transfers the chips from the presteaming vessel to the digester. In the digester, the wood chips are "cooked" using white liquor (a mixture of cooking chemicals including sodium hydroxide, sodium sulfide, sodium carbonate, and water) to solubilize the lignin in the wood chips. In the cooking process, methanol is produced. Following digestion of the lignin, the cooking chemicals are washed out of the pulp. A countercurrent multistage washing unit is utilized to minimize the carryover of chemicals with the pulp. The residual chemicals from the pulping process are called the weak black liquor. The black liquor contains sodium salts (hydroxide, sulfide, carbonate, chloride, sulfite, and sulfate), dissolved lignin, methanol, and water. In order to concentrate the weak black liquor, it is flashed twice before going to evaporator system. The steam generated in the first flash tank is used for presteaming operation, whereas the steam generated in the second flash tank is used in the chip bin unit. Vapors leaving the first flash tank (containing almost 90% of the methanol generated in the digester) join the vapors from the presteaming vessel and are fed to the primary condenser. In the primary condenser, a wastewater stream is generated. The noncondensables which contain most of the turpentine generated in the digester are fed to the secondary condenser where most of the gases are condensed and fed to the turpentine decanter. After recovering the turpentine in the decanter, the effluent is mixed with the wastewater generated in the primary condenser to form wastewater stream W_6. In the direct contact condenser, direct-contact cooling water is utilized to condense the noncondensables from the secondary condenser resulting in wastewater stream W_7. The liquor leaving the second flash tank (15–20% solids) is fed to a multiple-effect evaporators. The evaporation process results in the generation of a large amount of combined condensate which is classified as wastewater stream, W_{12}.

The water with the pulp is removed in two stages. In the first stage the pulp is concentrated by removing 300 ton/hr of water (W_{13}). This water is then filtered to remove

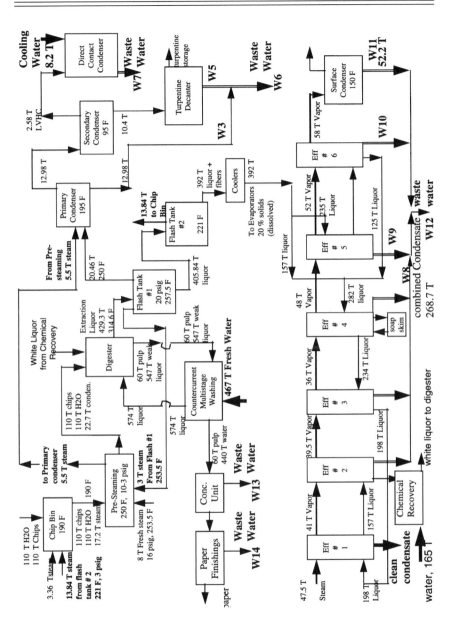

Figure 4.15 A Kraft pulping process (Hamad et al., 1995) (Basis: one hour, T refers to metric ton).

any suspended solids. In the second stage, the water is vaporized, condensed, and drained (W_{14}). All the wastewater streams (except W_{14}) are treated using biotreatment and then discharged to the river.

Characteristics of Wastewater Streams (Sources)

As can be seen from the process flowsheet, the following terminal wastewater streams are generated:
- Wastewater stream W_6 (22.68 ton/hr, 7,189 ppmw methanol). This stream consists of the underflow from the primary condenser W_3 (12.98 ton/hr, 419 ppmw methanol) and the underflow from the turpentine decanter W_5 (9.7 ton/hr, 16,248 ppmw methanol)
- Wastewater stream W_7 (10.78 ton/hr, 9900 ppmw methanol) resulting from (8.20 ton/hr of fresh water that is added to condense a vapor stream (2.58 ton/hr, 41,360 ppmw methanol)
- Combined condensate W_{12} (268.7 ton/hr, 114 ppmw methanol). This stream consists of combined condensate from second, third, and fourth evaporators (W_8, 116.5 ton/hr, 20 ppmw methanol), condensate from fifth evaporator (W_9, 48.0 ton/hr, 233 ppmw methanol), condensate from sixth evaporator (W_{10}, 52.0 ton/hr, 311 ppmw methanol), and condensate from surface condenser (W_{11}, 52.2 ton/hr, 20 ppmw methanol).
- Wastewater from the concentration and filtration units W_{13} (300 ton/hr, 30 ppmw methanol).
- Condensed wastewater in the paper-making process W_{14} (140 ton/hr, 15 ppmw methanol).

Sinks Characteristics

The following constraints on flowrates and compositions entering process sinks must be satisfied in any design:
- Pulp washing
    acceptable flowrate of wash water $= 467$ ton/hr
    acceptable composition of methanol in wash water ≤ 20 ppmw
- Chemical recovery
    165 ton/hr $\leq$ acceptable flowrate of water fed to chemical recovery ≤ 180 ton/hr
    acceptable composition of methanol in water fed to chemical recovery ≤ 20 ppmw
- Direct contact condenser
    acceptable flowrate of water fed to condenser $= 8.2$ ton/hr
    acceptable composition of methanol in water fed to condenser ≤ 10 ppmw
- River: Any stream discharged to the river should not have a methanol composition which exceeds 15 ppmw.

Interception Techniques

Several interception techniques should be considered to remove methanol. Air stripping is among these methods. The equilibrium relation for stripping from methanol aqueous streams may be approximated by the following expression

$$y = 0.38x,$$

where y and x are the mass compositions of methanol in water and air, respectively. A minimum allowable composition difference of 4,275 ppmw is recommended. Flowrate of air is determined as follows:

$$L = 0.5^* f^* G,$$

where L and G are the mass flowrates (kg/hr) of air and wastewater, respectively, and f is the fractional mass removal of methanol from water by stripping. The operating cost for air stripping is given by the following relationship:

$$\text{Operating Cost (\$/hr)} = 0.003^* L (\text{kg air/hr}).$$

This cost includes air compression and methanol condensation.

Biotreatment

The following information are available for the biotreatment facility:

acceptable methanol composition entering biotreatment $\leq 1,000$ ppmw
average outlet methanol composition $= 15$ ppmw
biotreatment operating cost $= 0.11^* M + 0.0013^* G,$

where M is the mass load (kg/hr) of methanol and G is the flowrate of wastewater (kg/hr).

(a) It is anticipated that recovering methanol from aqueous streams may provide methanol sales that are higher than recovery costs. Apply mass-integration techniques to determine which streams should be intercepted to recover methanol, the extent of recovery, and the optimum stream allocation. In addition to methanol recovery, outline the other benefits of your solution, such as reduced water usage and reduced wastewater discharge. **(Hints:** start with streams W_6 and W_7 and consider using an indirect contact condenser whose total annualized cost is $25,000/yr in place of the direct contact condenser).

(b) The company plans to build a new pulp and paper facility elsewhere. The capacity of the new process is the same as the one shown in figure. Develop a water allocation scheme for this grass root design that minimizes the total annualized cost of the water system including cost of any equipment (e.g., strippers, condensers) and treatment systems (e.g., biotreatment facility).

References

El-Halwagi, M. M., Hamad, A. A., and Garrison, G. W. (1996), "Synthesis of Waste Interception and Allocation Networks," Vol. 42, No. 11, pp. 3087–3101.

El-Halwagi, M. M., and Spriggs, H. D. (1996), "An Integrated Approach to Cost and Energy Efficient Pollution Prevention," *Fifth World Congress of Chemical Engineering*, Vol. III, pp. 344–349, San Diego.

Garrison, G. W., Hamad, A. A., and El-Halwagi, M. M. (1995), "Synthesis of Waste Interception Networks," AIChE Annual Meeting, Miami.

Garrison, G. W., Spriggs, H. D., and El-Halwagi, M. M. (1996), "A Global Approach to Integrating Environmental, Energy, Economic and Technological Objectives," *Fifth World Congress of Chemical Engineering*, Vol. I, pp. 675–680, San Diego.

Hamad, A. A., Garrison, G. W., Crabtree, E. W., and El-Halwagi, M. M. (1996), "Optimal Design of Hybrid Separation Systems for Waste Reduction," *Fifth World Congress of Chemical Engineers*, Vol. III, pp. 453–458, San Diego, California.

Hamad, A. A., Varma, V., El-Halwagi, M. M., and Krishnagopalan, G. (1995), "Systematic Integration of Source Reduction and Recycle Reuse for the Cost-Effective Compliance with the Cluster Rules," AIChE Annual Meeting, Miami.

CHAPTER FIVE

Synthesis of Mass-Exchange Networks—An Algebraic Approach

The graphical pinch analysis presented in Chapter Three provides the designer with a very useful tool that represents the global transfer of mass from the waste streams to the MSAs and determines performance targets such as MOC of the MSAs. Notwithstanding the usefulness of the pinch diagram, it is subject to the accuracy problems associated with any graphical approach. This is particularly true when there is a wide range of operating compositions for the waste and the lean streams. In such cases, an algebraic method is recommended. This chapter presents an algebraic procedure which yields results that are equivalent to those provided by the graphical pinch analysis. In addition, this chapter describes a systematic technique for matching waste-lean pairs and configuring MENs that realize the MOC solutions.

5.1 The Composition-Interval Diagram "CID"

The CID is a useful tool for insuring thermodynamic feasibility of mass exchange. On this diagram, $N_{sp} + 1$ corresponding composition scales are generated. First, a composition scale, y, for the waste streams is established. Then, Eq. (3.5) is employed to create N_{sp} corresponding composition scales for the process MSAs. On the CID, each process stream is represented as a vertical arrow whose tail corresponds to its supply composition while its head represents its target composition. Next, horizontal lines are drawn at the heads and tails of the arrows. These horizontal lines define a series of composition intervals. The number of intervals

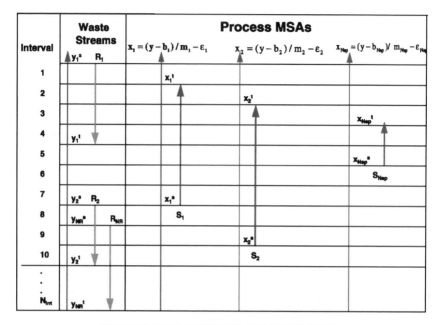

COMPOSITION-INTERVAL DIAGRAM (CID)

Figure 5.1 The composition interval diagram "CID".

is related to the number of process streams via

$$N_{int} \leq 2(N_R + N_{SP}) - 1, \qquad (5.1)$$

with the equality applying in cases where no heads on tails coincide. The composition intervals are numbered from top to bottom in an ascending order. The index k will be used to designate an interval with $k = 1$ being the uppermost interval and $k = N_{int}$ being the lowermost interval. Figure 5.1 provides a schematic representation of the CID. Within any interval, it is thermodynamically feasible to transfer mass from the waste streams to the MSAs. It is also feasible to transfer mass from a waste stream in an interval k to any MSA which lies in an interval $\bar{k}$ below it (i.e., $\bar{k} \geq k$).

5.2 Table of Exchangeable Loads "TEL"

The objective of constructing a TEL is to determine the mass-exchange loads of the process streams in each composition interval. The exchangeable load of the

ith waste stream which passes through the kth interval is defined as

$$W_{i,k}^R = G_i(y_{k-1} - y_k),\qquad(5.2)$$

where y_{k-1} and y_k are the waste-scale compositions of the transferrable species which respectively correspond to the top and the bottom lines defining the kth interval. On the other hand, the exchangeable load of the jth process MSA which passes through the kth interval is computed through the following expression

$$W_{j,k}^S = L_j^C(x_{j,k-1} - x_{j,k}),\qquad(5.3)$$

where $x_{j,k-1}$ and $x_{j,k}$ are the compositions on the jth lean-composition scale which respectively correspond to the higher and lower horizontal lines bounding the kth interval. Clearly, if a stream does not pass through an interval, its load within that interval is zero.

Having determined the individual loads of all process streams for all composition intervals, one can also obtain the collective loads of the waste and the lean streams. The collective load of the waste streams within the kth interval is calculated by summing up the individual loads of the waste streams that pass through that interval, i.e.

$$W_k^R = \sum_{i\ passes\ through\ interval\ k} W_{i,k}^R.\qquad(5.4)$$

Similarly, the collective load of the lean streams within the kth interval is evaluated as follows:

$$W_k^S = \sum_{j\ passes\ through\ interval\ k} W_{j,k}^S.\qquad(5.5)$$

We are now in a position to incorporate material balance into the synthesis procedure with the objective of allocating the pinch point as well as evaluating excess capacity of process MSAs and load to be removed by external MSAs. These aspects are assessed through the mass-exchange cascade diagram.

5.3 Mass-Exchange Cascade Diagram

As has been mentioned earlier, the CID generates a number N_{int} of composition intervals. Within each interval, it is thermodynamically as well as technically feasible to transfer a certain mass of the key pollutant from a waste stream to a lean stream. Furthermore, it is feasible to pass mass from a waste stream in an interval to any lean stream in a lower interval. Hence, for the kth composition interval, one can write the following component material balance for the key pollutant:

$$W_k^R + \delta_{k-1} - W_k^S = \delta_k,\qquad(5.6)$$

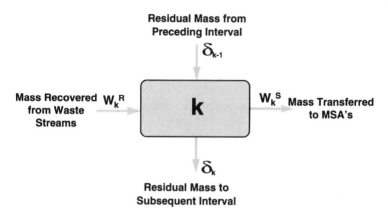

Figure 5.2 A pollutant material balance around a composition interval.

where δ_{k-1} and δ_k are the residual masses of the key pollutant entering and leaving the kth interval. Equation (5.6) indicates that the total mass input of the key component to the kth interval is due to collective load of the waste streams in that interval as well as the residual mass of the key component leaving the interval above it, δ_{k-1}. A total mass, W_k^S, of the key pollutant is transferred to the MSAs in the kth interval. Hence, a residual mass, δ_k, of the pollutant leaving the kth interval can be calculated via Eq. (5.6). This output residual also constitutes the influent residual to the subsequent interval. Figure 5.2 illustrates the component material balance for the key pollutant around the kth composition interval.

It is worth pointing out that δ_0 is zero since no waste streams exist above the first interval. In addition, thermodynamic feasibility is insured when all the δ_k's are nonnegative. Hence, a negative δ_k indicates that the capacity of the process lean streams at that level is greater than the load of the waste streams. The most negative δ_k corresponds to the excess capacity of the process MSAs in removing the pollutant. Therefore, this excess capacity of process MSAs should be reduced by lowering the flowrate and/or the outlet composition of one or more of the MSAs. After removing the excess capacity of MSAs, one can construct a revised TEL in which the flowrates and/or outlet compositions of the process MSAs have been adjusted. Consequently a revised cascade diagram can be generated. On the revised cascade diagram the location at which the residual mass is zero corresponds to the mass-exchange pinch composition. As expected, this location is the same as that with the most negative residual on the original cascade diagram. Since an overall material balance for the network must be realized, the residual mass leaving the lowest composition interval of the revised cascade diagram must be removed by external MSAs.

5.4 Example on Dephenolization of Aqueous Wastes

This is the same case study described by Section 3.8. Here, we will tackle the problem through the aforementioned algebraic technique. As has been previously described, the first step is to create the CID for the process streams as shown in Fig. 5.3. Then, we construct the TEL for the problem. This is shown in Table 5.1.

Next, the mass-exchange cascade diagram is generated. As can be seen in Fig. 5.4, the most negative residual mass is -0.0184 kg/s. This value corresponds to the excess capacity of process MSAs. It is worth noting that an identical result was obtained in Chapter Three through the pinch diagram. Such excess capacity can be removed by reducing the flowrates and/or outlet compositions of the process

Table 5.1
The TEL for the Dephenolization Example

Interval	Load of waste streams kg phenol/s			Load of process MSAs kg phenol/s		
	R_1	R_2	$R_1 + R_2$	S_1	S_2	$S_1 + S_2$
1	0.0052	–	0.0052	–	–	–
2	0.0308	–	0.0308	–	0.0303	0.0303
3	0.0040	–	0.0040	0.0050	0.0039	0.0089
4	0.0264	0.0132	0.0396	0.0330	0.0258	0.0588
5	0.0096	0.0048	0.0144	0.0120	–	0.0120
6	0.0040	0.0020	0.0060	–	–	–
7	–	0.0040	0.0040	–	–	–

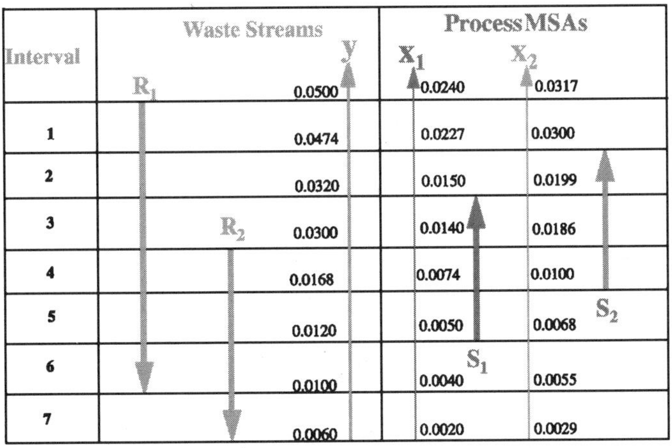

Figure 5.3 The CID for the dephenolization example.

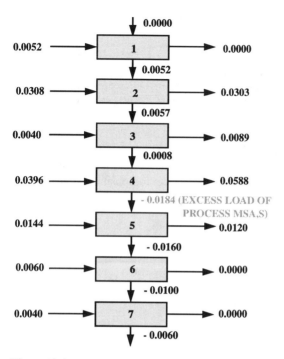

Figure 5.4 The cascade diagram for the dephenolization example.

MSAs. If we decide to eliminate this excess by decreasing the flowrate of S_2, the actual flowrate of S_2 should be 2.08 kg/s as has been calculated via Eq. (3.12). Using the adjusted flowrate of S_2, we can now construct the revised TEL for the problem as depicted by Table 5.2.

Table 5.2
The Revised TEL for the Dephenolization Example

Interval	Load of waste streams kg phenol/s			Load of process MSAs kg phenol/s		
	R_1	R_2	$R_1 + R_2$	S_1	S_2	$S_1 + S_2$
1	0.0052	–	0.0052	–	–	–
2	0.0308	–	0.0308	–	0.0210	0.0210
3	0.0040	–	0.0040	0.0050	0.0027	0.0077
4	0.0264	0.0132	0.0396	0.0330	0.0179	0.0509
5	0.0096	0.0048	0.0144	0.0120	–	0.0120
6	0.0040	0.0020	0.0060	–	–	–
7	–	0.0040	0.0040	–	–	–

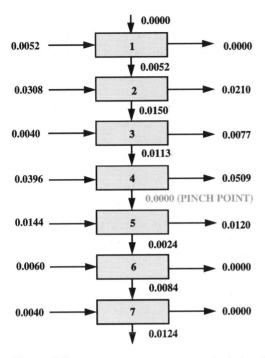

Figure 5.5 The revised cascade diagram for the dephenolization example.

Hence, the revised mass-exchange cascade diagram is generated as shown in Fig. 5.5. On this diagram, the residual mass leaving the fourth interval is zero. Therefore, the mass-exchange pinch is located on the line separating the fourth and the fifth intervals. As can be seen in Fig. 5.3, this location corresponds to a set of corresponding composition scales $(y, x_1, x_2) = (0.0168, 0.0074, 0.0100)$. Furthermore, Fig. 5.5 shows the residual mass leaving the bottom interval being 0.0124 kg/s. This value is the amount of pollutant to be removed by external MSAs. Again, a similar result has been obtained by the pinch diagram (Fig. 3.17).

5.5 Synthesis of MENs with Minimum Number of Exchangers

As has been mentioned in Section 3.4, the targeting approach adopted for synthesizing MENs attempts to first minimize the cost of MSAs by identifying the flowrates and outlet compositions of MSAs which yield minimum operating cost "MOC". This target has been tackled in Chapter Three as well as in the foregoing sections in this chapter. The second step in the synthesis procedure is to minimize the number of exchangers which can realize the MOC solution. Owing to the existence of a pinch which decomposes the problem into two subproblems; one above the pinch and one below the pinch, the minimum number of mass exchangers compatible with a MOC solution, U_{MOC}, can be obtained by applying Eq. (3.2) to each subproblem separately, i.e.,

$$U_{MOC} = U_{MOC, \text{above pinch}} + U_{MOC, \text{below pinch}}, \tag{5.7a}$$

where

$$U_{MOC, \text{above pinch}} = N_{R, \text{above pinch}} + N_{S, \text{above pinch}} - N_{i, \text{above pinch}} \tag{5.7b}$$

and

$$U_{MOC, \text{below pinch}} = N_{R, \text{below pinch}} + N_{S, \text{below pinch}} - N_{i, \text{below pinch}} \tag{5.7c}$$

Having determined U_{MOC}, we should then proceed to match the pairs of waste and lean streams. In the sequel, it will be shown that matching has to start from the pinch and must satisfy a number of feasibility criteria.

5.6 Feasibility Criteria at the Pinch

In order to guarantee the minimum cost of MSAs, no mass should be transferred across the pinch. Therefore, by starting stream matching at the pinch the designer avoids any situation which may later result in transferring mass through the pinch. Moreover, at the pinch all matches feature a driving force (between operating and equilibrium lines) equal to the minimum allowable composition difference, ε_j. Hence, since the pinch represents the most thermodynamically-constrained region for design, the number of feasible matches in this region is severely limited. Thus, when the synthesis is started at the pinch, the freedom of design choices at later steps will not be prejudiced.

The above discussion clearly shows that the synthesis of a MEN should start at the pinch and proceed in two directions separately: the rich and the lean ends. To facilitate the design procedure both above and below the pinch, we next establish feasibility criteria for the characterization of stream matches at the pinch. These criteria identify the essential matches or topology options at the pinch. Such matches will be referred to as "pinch matches" or "pinch exchangers." The feasibility criteria will also inform the designer whether or not stream splitting is required at the pinch. Once away from the pinch, these feasibility criteria do not have to be considered, and matching of the waste and the lean streams becomes a relatively simple task. In order to identify the feasible pinch topologies, the following two feasibility criteria will be applied to the stream data:

(i) Stream population
(ii) Operating line versus equilibrium line.

5.6.1 Stream Population

Let us first consider the case above the pinch. In a MOC design, any mass exchanger immediately above the pinch will operate with the minimum allowable composition difference at the pinch side. Since it is required to bring all the waste streams at the pinch to their pinch composition, any waste stream leaving a pinch exchanger can only operate against a lean stream at its pinch composition. Therefore, for each pinch match, at least one lean stream (or branch) has to exist per each waste stream. In other words, for a MOC design, the following inequality must apply at the rich end of the pinch

$$N_{ra} \leq N_{la}, \tag{5.8a}$$

where N_{ra} is the number of waste (rich) streams or branches immediately above the pinch, and N_{la} is the number of lean streams or branches immediately above the pinch. If the above inequality does not hold for the stream data, one or more of the lean streams will have to be split.

Conversely, immediately below the pinch, each lean stream has to be brought to its pinch composition. At this composition, any lean stream can only operate against a waste stream at its pinch composition or higher. Since a MOC design does not permit the transfer of mass across the pinch, each lean stream immediately below the pinch will require the existence of at least one waste stream (or branch) at the pinch composition.

Therefore, immediately below the pinch, the following criteria must be satisfied:

$$N_{lb} \leq N_{rb} \tag{5.8b}$$

where N_{lb} is the number of lean streams or branches immediately below the pinch, and N_{rb} is the number of waste (rich) streams or branches immediately below the pinch. Again, splitting of one or more of the waste streams may be necessary to realize the above inequality.

5.6.2 Operating Line versus Equilibrium Line

Consider the mass exchanger shown in Fig. 5.6. The lean end of this exchanger is immediately above the pinch. A component material balance for the pollutant around the exchanger can be written as

$$G_i \left(y_i^{in} - y_i^{pinch} \right) = L_j \left(x_j^{out} - x_j^{pinch} \right), \tag{5.9}$$

but, at the pinch,

$$y_i^{pinch} = m_j \left(x_j^{pinch} + \varepsilon_j \right) + b_j. \tag{5.10}$$

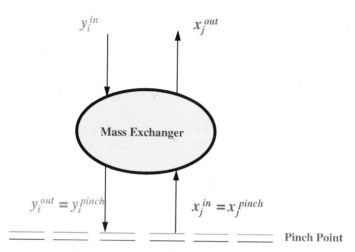

Figure 5.6 A mass exchanger immediately above the pinch.

In order to ensure thermodynamic feasibility at the rich end of the exchanger, the following inequality must hold

$$y_i^{\text{in}} \geq m_j \left(x_j^{\text{out}} + \varepsilon_j\right) + b_j. \tag{5.11}$$

Substituting from Eqs. (5.10) and (5.11) into Eq. (5.9), one gets

$$G_i \left[m_j \left(x_j^{\text{out}} + \varepsilon_j\right) + b_j - m_j \left(x_j^{\text{pinch}} + \varepsilon_j\right) - b_j\right] \leq L_j \left(x_j^{\text{out}} - x_j^{\text{pinch}}\right),$$

and hence

$$\frac{L_j}{m_j} \geq G_i. \tag{5.12a}$$

This is the feasibility criterion for matching a pair of streams (i, j) immediately above the pinch. By recalling that the slope of the operating line is L_j/G_i, one can phrase Eq. (5.12) as follows: In order for a match immediately above the pinch to be feasible, the slope of the operating line should be greater than or equal to the slope of the equilibrium line. On the other hand, one can similarly show that the feasibility criterion for matching a pair of streams (i, j) immediately below the pinch is given by

$$\frac{L_j}{m_j} \leq G_i. \tag{5.12b}$$

Once again, stream splitting may be required to guarantee that inequality (5.12a) or (5.12b) is realized for each pinch match. It should also be emphasized that the feasibility criteria (Eqs. 5.8 and 5.12) should be fulfilled only at the pinch. Once the pinch matches are identified, it generally becomes a simple task to complete the network design. Moreover, the designer always has the freedom to violate these feasibility criteria at the expense of increasing the cost of external MSAs beyond the MOC requirement.

5.7 Network Synthesis

The feasibility criteria described by Eqs. (5.8) and (5.12) can be employed to synthesize a MEN which has the minimum number of exchangers that satisfy the MOC solution. The following representation will be used to illustrate the network structure graphically. Waste streams are represented by vertical arrows running at the left of the diagram. Compositions (expressed as weight ratios of the key component in each stream) are placed next to the corresponding arrow. A match between two streams is indicated by placing a pair of circles on each of the streams and connecting them by a line. Mass-transfer loads of the key component for each exchanger are noted in appropriate units (e.g., kg pollutant/s) inside the circles. The pinch is represented by two horizontal dotted lines.

In order to demonstrate the applicability of the foregoing feasibility criteria, we now revisit the dephenolization case study. As has been discussed earlier, the synthesis ought to start at the pinch and proceed in two directions; the rich and the lean ends. Let's begin with the rich end of the problem. Above the pinch, we have two waste streams and two MSAs. Hence, minimum number of exchangers above the pinch can be calculated according to Eq. (5.7b) as

$$U_{\text{MOC, above pinch}} = 2 + 2 - 1 = 3 \text{ exchangers} \tag{5.13}$$

Immediately above the pinch, the number of rich streams is equal to the number of the MSAs, thus the feasibility criterion given by Eq. (5.8a) is satisfied. The second feasibility criterion (Eq. 5.12a) can be checked through Fig. 5.7. By comparing the values of $\frac{L_j}{m_j}$ with G_i for each potential pinch match, one can readily deduce that it is feasible to match S_1 with either R_1 or R_2 immediately above the pinch. Nonetheless, while it is possible to match S_2 with R_2, it is infeasible to pair S_2 with R_1 immediately above the pinch. Therefore, one can match S_1 with R_1 and S_2 with R_2 as rich-end pinch exchangers.

When two streams are paired, the exchangeable mass is the lower of the two loads of the streams. For instance, the mass-exchange loads of R_1 and S_1 are 0.0664 and 0.0380 kg/s, respectively. Hence, the mass exchanged from R_1 to S_1 is 0.0380 kg/s. Owing to this match, the capacity of S_1 above the pinch has been completely exhausted and S_1 may now be eliminated from any further consideration in the rich-end subproblem. Similarly, 0.0132 kg/s of phenol will be transferred from R_2 to S_2 thereby fulfilling the required mass-exchange duty for R_2 above the pinch. We are now left with two streams only above the pinch

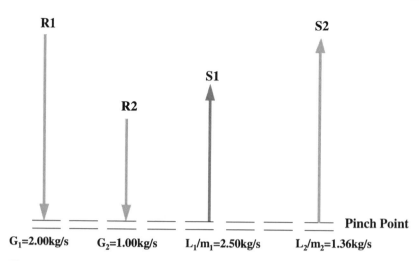

$G_1=2.00kg/s$ $G_2=1.00kg/s$ $L_1/m_1=2.50kg/s$ $L_2/m_2=1.36kg/s$

Figure 5.7 Feasibility criteria above the pinch for the dephenolization example.

(R_1 and S_2). The remaining load for R_1 is $0.0664 - 0.0380 = 0.0284$ kg/s. Also, the removal capacity left in S_2 is $0.0416 - 0.0132 = 0.0284$ kg/s. As expected, both remaining loads of R_1 and S_2 above the pinch are equal. This is attributed to the fact that no mass is passed through the pinch. Hence, material balance must be satisfied above the pinch. The two streams (R_1 and S_2) are, therefore, matched and the synthesis subproblem above the pinch is completed. This rich-end design is shown in Fig. 5.8. The intermediate compositions can be calculated through component material balances. For instance, the composition of S_2 leaving its match with R_2 and entering is match with R_1, $x_2^{\text{intermediate}}$, can be calculated via a material balance around the R_2-S_2 exchanger, i.e.,

$$x_2^{\text{intermediate}} = 0.0100 + \frac{0.0132}{2.08} = 0.0164 \qquad (5.14a)$$

or a material balance around the R_1-S_2 exchanger:

$$x_2^{\text{intermediate}} = 0.0300 - \frac{0.0284}{2.08} = 0.0164 \qquad (5.14b)$$

Having completed the design above the pinch, we can now move to the problem below the pinch. Figure 5.9 illustrates the streams below the pinch. It is worth noting that immediately below the pinch, only streams R_1, R_2 and S_1 exist. Stream S_3 does not reach the pinch point and, hence, will not be considered when the

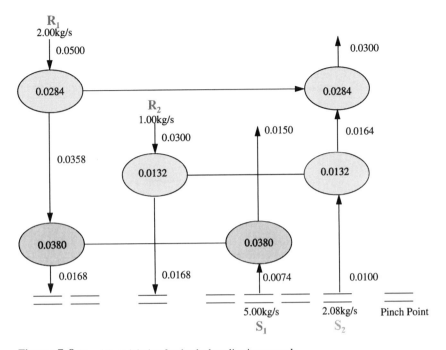

Figure 5.8 The rich-end design for the dephenolization example.

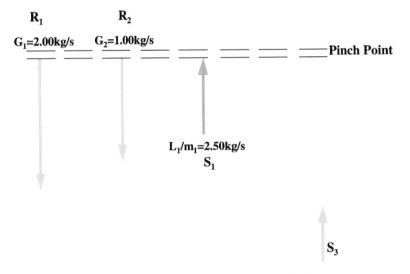

Figure 5.9 Feasibility criteria below the pinch for the dephenolization example.

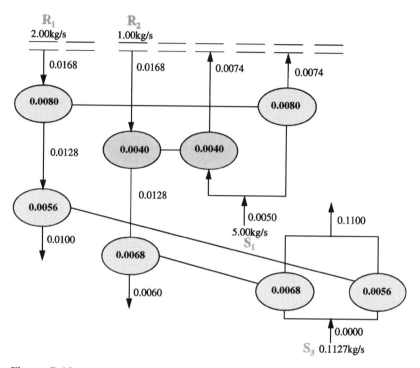

Figure 5.10 A lean-end design for the dephenolization example.

feasibility criteria of matching streams at the pinch are applied. Since N_{rb} is 2 and N_{lb} is 1, inequality (5.8b) is satisfied. Next, the second feasibility criterion (5.12b) is examined. As can be seen in Fig. 5.8, S_1 cannot be matched with either R_1 or R_2 since L_1/m_1 is greater than G_1 and G_2. Hence, S_1 must be split into two branches; one to be matched with R_1 and the other to be paired with R_2. There are infinite number of ways through which L_1 can be split so as to satisfy Eq. (5.12b). Let us arbitrarily split L_1 in the same ratio of G_1 to G_2, i.e., to 3.33 and 1.67 kg/s. This split realizes the inequality (2.12b) since $3.33/2 < 2$ and $1.67/2 < 1$.

The remaining loads of R_1 and R_2 can now be eliminated by S_3 (activated carbon). Several configurations can be envisioned for S_3. These structures include a split design (Fig. 5.10), a serial design in which S_3 if first matched with R_1 (Fig. 5.11), and a serial design in which S_3 is first matched with R_2 (Fig. 5.12). It is worth pointing out that the number of exchangers below the pinch is four which is one more than $U_{MOC,\,below\,the\,pinch}$. Once again, $U_{MOC,\,below\,the\,pinch}$ is just a lower bound on the number of exchangers and does not have to be exactly realized.

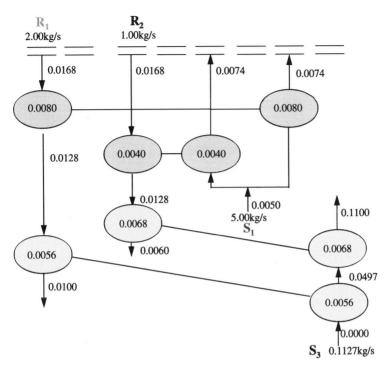

Figure 5.11 A lean-end design for the dephenolization example.

It is now possible to generate a MOC network by combining the rich-end design with a lean-end design. For instance, if Figs. 5.8 and 5.10 are merged, the MOC MEN shown in Fig. 5.13 is obtained.

5.8 Trading Off Fixed versus Operating Costs Using Mass-Load Paths

Section 3.11 discussed two common methods for trading off fixed versus operating costs. These two methods involve varying the minimum allowable composition differences and assessing the mixing of waste streams. A third method for trading off fixed versus operating costs is the use of "mass-load paths" (El-Halwagi and Manousiouthakis, 1989). A mass-load path is a continuous connection which starts with an external MSA and concludes with a process MSA. By shifting the loads along a path, one can add an excess amount of external MSA to replace an equivalent amount of process MSA. The end result of this procedure is the elimination of exchangers at the expense of incurring additional operating cost. For instance,

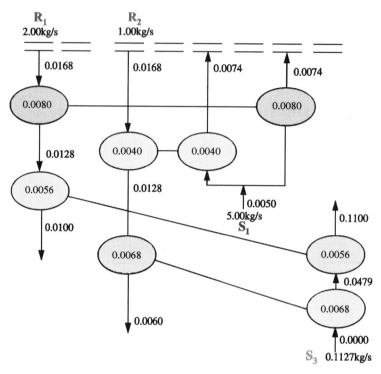

Figure 5.12 A lean-end design for the dephenolization example.

in Fig. 5.13, the path S_3-R_2-S_1 can be used to shift a load of 0.0040 kg/s from S_1 to S_3. This load shift will lead to the elimination of the exchanger pairing R_2 with S_1 below the pinch. Therefore, the flowrate and/or outlet composition of S_1 should be reduced to remove the unused capacity of 0.004 kg phenol/s. For instance, if the outlet composition of S_1 is kept unchanged, the flowrate of S_1 should be adjusted to 4.60 kg/s. An additional benefit of load shifting is that the two exchangers matching S_1 with R_1 above and below the pinch can now be combined into a single exchanger. However, the flowrate of S_3 will have to be increased by $0.0040/(0.1100 - 0) = 0.0364$ kg/s. This step incurs an additional operating cost which can be calculated as follow:

Additional annual operating cost due to increase in flowrate of S_3

$$= \left(0.0364 \frac{kg \text{ activated carbon}}{s} \right) \left(\frac{\$\ 0.081}{kg \text{ activated carbon}} \right)$$

$$\times \left(3600 \frac{s}{hr} \right) \left(\frac{8760\ hr}{yr} \right) \approx \$\ 93 \times 10^3 / yr. \tag{5.15a}$$

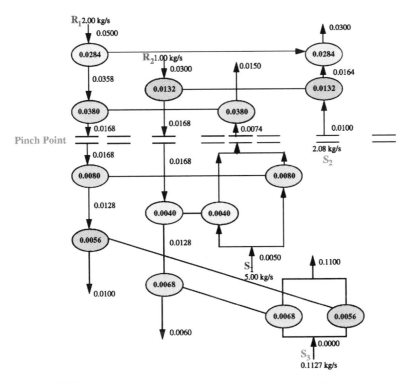

Figure 5.13 A complete MOC network for the dephenolization example.

or

Additional annual operating cost due to increase in flowrate of S_3

$$= \left(0.004\frac{kg\ phenol}{s}\right)\left(\frac{\$\ 0.737}{kg\ phenol\ removed}\right)\left(3600\frac{s}{hr}\right)\left(\frac{8760\ hr}{yr}\right)$$

$$\approx \$\ 93 \times 10^3/yr. \tag{5.15b}$$

The designer should compare this additional cost with the fixed-cost savings accomplished by eliminating two exchangers (due to the removal of the exchanger matching R_2 with S_1 below the pinch and the combination of the two exchangers matching R_1 with S_1 into a single unit). This comparison provides the basis for determining whether or not the mass-load path should be employed. Figure 5.14 illustrates the network configuration after the mass-load path is used to eliminate two exchangers.

Three additional mass-load paths (S_3-R_2-S_2, S_3-R_1-S_1, S_3-R_1-S_2) can be employed to shift removal duties from the process MSAs to the external MSA until all the waste load (0.104 kg phenol/s) is removed by the external MSA

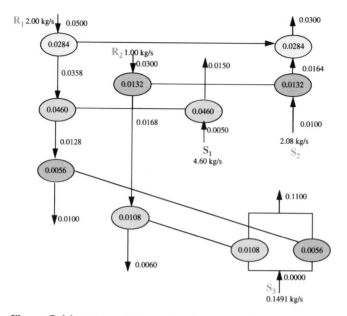

Figure 5.14 Reduction of the number of exchangers using a mass-load path.

(Fig. 5.15). In this case, the operating cost of the system is:

Annual operating cost

$$= \left(0.104 \frac{kg\ phenol}{s} \right) \left(\frac{\$\ 0.737}{kg\ phenol} \right) \left(3600 \frac{s}{hr} \right) \left(\frac{8760\ hr}{yr} \right)$$

$$\approx \$\ 2.417\ million/yr. \tag{5.16}$$

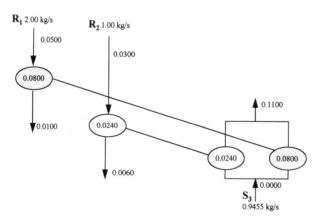

Figure 5.15 Elimination of process MSAs using mass-load paths.

By comparing Eqs. (3.29) and (5.16), we notice that the application of the mass-load paths has resulted in an increase of the operating cost of $2.129 million/yr. On the other hand, the number of mass exchangers has dropped from seven to two. The extent to which the fixed cost of the mass exchangers has decreased determines whether or not this increase in operating cost is acceptable.

Problems

5.1 Using an algebraic procedure, synthesize an optimal MEN for the benzene recovery example described in Section 3.7 (Example 3.1).

5.2 Resolve the dephenolization example presented in this chapter for the case when the two waste streams are allowed to mix.

5.3 Synthesize an MEN featuring the minimum number of units that realize the MOC solution for Problem 3.1.

5.4 Using the cascade diagram, solve Problem 3.3. Also, synthesize an MEN which has the minimum number of units satisfying the MOC solution. Employ the mass-load paths to reduce the number of mass exchangers at the expense of increasing operating cost.

5.5 The techniques presented in this chapter can be generalized to tackle MENs with multiple pollutants. Consider the COG sweetening process addressed by the previous problem. Carbon dioxide often exists in COG in relatively large concentrations. Therefore, partial removal of CO_2 is sometimes desirable to improve the heating value of COG, and almost complete removal of CO_2 is required for gases undergoing low-temperature processing. The data for the CO_2-laden rich and lean streams are given in Table 5.3. Synthesize an MOC MEN which removes hydrogen sulfide and carbon dioxide simultaneously. Also, use mass-load paths to trade off fixed versus operation costs. **Hint:** See El-Halwagi and Manousiouthakis (1989 a,b).

Table 5.3
Stream Data for the COG-Sweetening Problem Rich Stream MSAs

Rich streams				MSAs						
Stream	G_i (kg/s)	Supply mass fraction of CO_2	Target mass fraction of CO_2	Stream	L_j^c kg/s	Supply mass fraction of CO_2	Target mass fraction of CO_2	m_j for CO_2	b_j for CO_2	c_j $/kg
R_1	0.90	0.0600	0.0050	S_1	2.3	0.0100	?	0.35	0.000	0.00
R_2	0.10	0.1150	0.0100	S_2	∞	0.0003	?	0.58	0.000	0.10

5.6 Use the algebraic procedure to synthesize an MEN which has the minimum number of units satisfying the MOC solution for Problem 3.4. **Hint:** Due to the absence of a pinch, there is no need to examine the matching feasibility criteria given by Eqs. (5.8) and (5.12). All you need is to check the practical feasibility Eq. (3.5) at the two ends of each exchanger.

Symbols

b_j	intercept of equilibrium line of the jth MSA
G_i	flowrate of the ith waste stream
i	index of waste streams
j	index of MSAs
k	index of composition intervals
L_j	flowrate of the jth MSA
L_j^C	upper bound on available flowrate of the jth MSA
m_j	slope of equilibrium line for the jth MSA
$N_{i,\,above\,pinch}$	number of independent synthesis problems above the pinch
$N_{i,\,below\,pinch}$	number of independent synthesis problems below the pinch
N_{int}	number of composition intervals
N_{ra}	number of rich streams immediately above the pinch
$N_{R,\,above\,pinch}$	number of rich streams above the pinch
N_{rb}	number of rich streams immediately below the pinch
$N_{R,\,below\,pinch}$	number of rich streams below the pinch
N_{sa}	number of lean streams immediately above the pinch
$N_{S,\,above\,pinch}$	number of lean streams above the pinch
N_{sb}	number of lean streams immediately below the pinch
$N_{S,\,below\,pinch}$	number of lean streams below the pinch
N_{SP}	number of process MSAs
R_i	the ith waste stream
S_j	the jth lean stream
U_{MOC}	minimum number of mass exchangers compatible with MOC, defined by Eq. (5.7a)
$U_{MOC,\,above\,pinch}$	minimum number of mass exchangers satisfying MOC above pinch, defined by Eq. (5.7b)
$U_{MOC,\,below\,pinch}$	minimum number of mass exchangers satisfying MOC below pinch, defined by Eq. (5.7c)
$W_{i,k}^R$	the exchangeable lead of the ith waste stream which passes through the kth interval as defined by Eq. (5.2)
$W_{j,k}^S$	the exchangeable load of the jth MSA passing through the kth interval as defined by Eq. (5.3)
W_k^R	the collective exchangeable load of the waste streams in interval k as defined by Eq. (5.4)
W_k^S	the collective exchangeable load of the MSAs in interval k as defined by Eq. (5.5)
$x_{j,k-1}$	composition of key component in the jth MSA at the upper horizontal line defining the kth interval

$x_{j,k}$	composition of key component in the jth MSA at the lower horizontal line defining the kth interval
x_j^{pinch}	composition of key component in the jth MSA at the pinch
x_j^{out}	composition of key component in the jth MSA leaving a pinch exchanger
y_i^{in}	composition of key component in the ith waste stream entering a pinch exchanger
y_i^{pinch}	composition of key component in the ith waste stream at the pinch
y_{k-1}	composition of key component in the ith waste stream at the upper horizontal line defining the kth interval
y_k	composition of key component in the ith waste stream at the lower horizontal line defining the kth interval

References

El-Halwagi, M. M., and Manousiouthakis, V. (1989a). Synthesis of mass exchange networks. *AIChE J.*, **35**(8), 1233–1244.

El-Halwagi, M. M., and Manousiouthakis, V. (1989b). Design and analysis of mass exchange networks with multicomponent targets, AIChE Annual Meeting, San Francisco.

CHAPTER SIX

Synthesis of Mass-Exchange Networks: A Mathematical Programming Approach

As has been discussed in Chapter One, mathematical programming (or optimization) is a powerful tool for process integration. For an overview of optimization and its application in pollution prevention, the reader is referred to El-Halwagi (1995). In this chapter, it will be shown how optimization techniques enable the designer to:

- Simultaneously screen all MSAs even when there are no process MSAs
- Determine the MOC solution and locate the mass-exchange pinch point
- Determine the best outlet composition for each MSA
- Construct a network of mass exchangers which has the least number of units that realize the MOC solution.

6.1 Generalization of the Composition Interval Diagram

The notion of a CID has been previously discussed in Section 5.1. This notion will now be generalized by incorporating external MSAs. In the generalized CID, $N_S + 1$ composition scales are created. First, a single composition scale, y, is established for the waste streams. Next, Eq. (3.5) is utilized to generate N_S corresponding composition scales (N_{SP} for process MSAs and N_{SE} for external MSAs). The locations corresponding to the supply and target compositions of the streams determine a sequence of composition intervals. The number of these intervals depends on the number of streams through the following inequality

$$N_{int} \leq 2(N_R + N_S) - 1. \tag{6.1}$$

The construction of the CID allows the evaluation of exchangeable loads for each stream in each composition interval. Hence, one can create a TEL for the waste streams in which the exchangeable load of the ith waste stream within the kth interval is defined as

$$W_{i,k}^R = G_i(y_{k-1} - y_k) \qquad (6.2)$$

when stream i passes through interval k, and

$$W_{i,k}^R = 0 \qquad (6.3)$$

when stream i does not pass through interval k. The collective load of the waste streams within interval k, W_k^R, can be computed by summing up the individual loads of the waste streams that pass through that interval, i.e.,

$$W_k^R = \sum_{i \ passes \ through \ interval \ k} W_{i,k}^R \qquad (6.4)$$

On the other hand, since the flowrate of each MSA is unknown, exact capacities of MSAs cannot be evaluated. Instead, one can create a ***TEL per unit mass of the MSAs*** for the lean streams. In this table, the exchangeable load *per unit mass of the MSA* is determined as follows:

$$w_{j,k}^S = x_{j,k-1} - x_{j,k} \qquad (6.5)$$

for the jth MSA passing through interval k, and

$$w_{j,k}^S = 0 \qquad (6.6)$$

when the jth MSA does not pass through the kth interval.

6.2 Problem Formulation

Section 5.3 has presented a technique for evaluating thermodynamically-feasible material balances among process streams by using the mass-exchange cascade diagram. This technique will now be generalized to include external MSAs. Once again the objective is to minimize the cost of MSAs which can remove the pollutant from the waste streams in a thermodynamically-feasible manner. Since the flowrates of the MSAs are not known, the objective function as well as the material balances around composition intervals have to be written in terms of these flowrates. The solution of the optimization program determines the optimal flowrate of each MSA. Hence, the task of identifying the MOC of the problem can be formulated through the following optimization program (El-Halwagi and Manousiouthakis, 1990a):

$$min \sum_{j=1}^{N_S} C_j L_j \qquad (P6.1)$$

subject to

$$\delta_k - \delta_{k-1} + \sum_{j \text{ passes through interval } k} L_j w_{j,k}^S = W_k^R \quad k = 1,2,\ldots,N_{int}$$

$$L_j \geq 0, \quad j = 1, 2, \ldots, N_S$$

$$L_j \leq L_j^c, \quad j = 1, 2, \ldots, N_S$$

$$\delta_0 = 0,$$

$$\delta_{N_{int}} = 0,$$

$$\delta_k \geq 0 \quad k = 1, 2, \ldots, N_{int} - 1.$$

The above program (P6.1) is a linear program that seeks to minimize the objective function of the operating cost of MSAs where C_j is the cost of the jth MSA ($/kg of recirculating MSA, including regeneration and makeup costs) and L_j is the flowrate of the jth MSA. The first set of constraints represents successive material balances around each composition interval where δ_{k-1} and δ_k are the residual masses of the key pollutant entering and leaving the kth interval. The second and third sets of constraints guarantee that the optimal flowrate of each MSA is nonnegative and is less than the total available quantity of that lean stream. The fourth and fifth constraints ensure that the overall material balance for the problem is realized. Finally, the last set of constraints enables the waste streams to pass the mass of the pollutant downwards if it does not fully exchange it with the MSAs in a given interval. This transfer of residual loads is thermodynamically feasible owing to the way in which the CID has been constructed.

The solution of program (P6.1) yields the optimal values of all the L_j's ($j = 1, 2, \ldots, N_S$) and the residual mass-exchange loads δ_k's ($k = 1, 2, \ldots, N_{int} - 1$). The location of any pinch point between two consecutive intervals, k and $k + 1$, is indicated when the residual mass-exchange load δ_k vanishes. This is a *generalization of the concept of a mass-exchange pinch point* discussed in Section 3.6. Since the plant may not involve the use of any process MSAs, external MSAs can indeed be used above the pinch to obtain an MOC solution. However, the pinch point still maintains its significance as the most thermodynamically constrained region of the network at which all mass transfer duties take place with driving forces equal to the minimum allowable composition differences.

6.3 The Dephenolization Example Revisited

The dephenolization problem was described in Section 3.2. The data for the waste and the lean streams are summarized by Tables 6.1 and 6.2.

Table 6.1
Data for Waste Streams in Dephenolization Example

Stream	Description	Flowrate G_i (kg/s)	Supply composition y_i^s	Target composition y_i^t
R_1	Condensate from first stripper	2	0.050	0.010
R_1	Condensate from second stripper	1	0.030	0.006

The first step in determining the MOC is to construct the CID for the problem to represent the waste streams along with the process and external MSAs. The CID is shown in Fig. (6.1) for the case when the minimum allowable composition differences are 0.001. Hence, one can evaluate the exchangeable loads for the two waste streams over each composition interval. These loads are calculated through Eqs. (6.2) and (6.3). The results are illustrated by Table 6.3.

Next, using Eqs. (6.5) and (6.6), the TEL for the lean streams per unit mass of the MSA is created. These loads are depicted in Table 6.4.

We are now in a position to formulate the problem of minimizing the cost of MSAs. By adopting the linear-programming formulation (P6.1), one can write the following optimization program:

$$\min 0.081L_3 + 0.214L_4 + 0.060L_5 \qquad \text{(P6.2)}$$

Table 6.2
Data for MSAs in Dephenolization Example

Stream	Description	Upper bound on flowrate L_j^c (kg/s)	Supply composition x_j^s	Target composition x_j^t	Equilibrium distribution coefficient $m_j = y/x_j$	Cost C_j ($/kg of recirculation MSA)
S_1	Gas oil	5	0.005	0.015	2.00	0.000
S_2	Lube oil	3	0.010	0.030	1.53	0.000
S_3	Activated carbon	∞	0.000	0.110	0.02	0.081
S_4	Ion-exchange resin	∞	0.000	0.186	0.09	0.214
S_5	Air	∞	0.000	0.029	0.04	0.060

Table 6.3
TEL for Waste Streams

	Load of waste streams (kg phenol/s)		
Interval	R_1	R_2	$R_1 + R_2$
1	0.0052	–	0.0052
2	0.0308	–	0.0308
3	0.0040	–	0.0040
4	0.0264	0.0132	0.0396
5	0.0096	0.0048	0.0144
6	0.0040	0.0020	0.0060
7	–	0.0040	0.0040
8	–	–	–
9	–	–	–
10	–	–	–
11	–	–	–
12	–	–	–

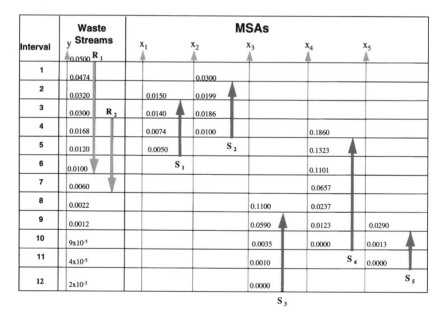

Figure 6.1 CID for dephenolization example.

Table 6.4

The TEL (kg Phenol/kg MSA) for the MSAs

Interval	S_1	S_2	S_3	S_4	S_5
	Capacity of lean streams per unit mass of MSA (kg phenol/kg MSA)				
1	–	–	–	–	–
2	–	0.0101	–	–	–
3	0.0010	0.0013	–	–	–
4	0.0066	0.0086	–	–	–
5	0.0024	–	–	0.0537	–
6	–	–	–	0.0222	–
7	–	–	–	0.0444	–
8	–	–	–	0.0420	–
9	–	–	0.0510	0.0114	–
10	–	–	0.0555	0.0123	0.0277
11	–	–	0.0025	–	0.0013
12	–	–	0.0010	–	–

subject to

$$\delta_1 = 0.0052$$

$$\delta_2 - \delta_1 + 0.0101L_2 = 0.0308$$

$$\delta_3 - \delta_2 + 0.0010L_1 + 0.0013L_2 = 0.0040$$

$$\delta_4 - \delta_3 + 0.0066L_1 + 0.0086L_2 = 0.0396$$

$$\delta_5 - \delta_4 + 0.0024L_1 + 0.0537L_4 = 0.0144$$

$$\delta_6 - \delta_5 + 0.0222L_4 = 0.0060$$

$$\delta_7 - \delta_6 + 0.0444L_4 = 0.0040$$

$$\delta_8 - \delta_7 + 0.0420L_4 = 0.0000$$

$$\delta_9 - \delta_8 + 0.0510L_3 + 0.0114L_4 = 0.0000$$

$$\delta_{10} - \delta_9 + 0.0555L_3 + 0.0123L_4 + 0.0277L_5 = 0.000$$

$$\delta_{11} - \delta_{10} + 0.0025L_3 + 0.0013L_5 = 0.0000$$

$$-\delta_{11} + 0.0010L_3 = 0.0000$$

$$\delta_k \geq 0, \quad k = 1, 2, \ldots, 11$$

$$L_j \geq 0, \quad j = 1, 2, \ldots, 5$$

$$L_1 \leq 5,$$

$$L_2 \leq 3.$$

In terms of LINGO input, program P6.2 can be written as:

```
model:
min = 0.081*L3 + 0.214*L4 + 0.060*L5;
delta1 = 0.0052;
delta2 - delta1 + 0.0101*L2 = 0.0308;
delta3 - delta2 + 0.001*L1 + 0.0013*L2 = 0.0040;
delta4 - delta3 + 0.0066*L1 + 0.0086*L2 = 0.0396;
delta5 - delta4 + 0.0024*L1 + 0.0537*L4 = 0.0144;
delta6 - delta5 + 0.0222*L4 = 0.0060;
delta7 - delta6 + 0.0444*L4 = 0.0040;
delta8 - delta7 + 0.0420*L4 = 0.0000;
delta9 - delta8 + 0.051*L3 + 0.0114*L4 = 0.000;
delta10 - delta9 + 0.0555*L3 + 0.0123*L4 + 0.0277*L5
= 0.000;
delta11 - delta10 + 0.0025*L3 + 0.0013*L5 = 0.0000;
-delta11 + 0.0010*L3 = 0.000;
delta1 >= 0.0;
delta2 >= 0.0;
delta3 >= 0.0;
delta4 >= 0.0;
delta5 >= 0.0;
delta6 >= 0.0;
delta7 >= 0.0;
delta8 >= 0.0;
delta9 >= 0.0;
delta10 >= 0.0;
delta11 >= 0.0;
L1 >= 0.0;
L2 >= 0.0;
L3 >= 0.0;
L4 >= 0.0;
L5 >= 0.0;
L1 <= 5.0;
L2 <= 3.0;
end
```

The following is the solution report generated by LINGO:

```
Optimal solution found at step:      2
Objective value:             0.9130909E-02
```

Variable	Value
L3	0.1127273
L4	0.0000000E+00
L5	0.0000000E+00
DELTA1	0.5200000E-02
DELTA2	0.1499200E-01
L2	2.080000
DELTA3	0.1128800E-01
L1	5.000000
DELTA4	0.0000000E+00
DELTA5	0.2399999E-02
DELTA6	0.8399999E-02
DELTA7	0.1240000E-01
DELTA8	0.1240000E-01
DELTA9	0.6650909E-02
DELTA10	0.3945454E-03
DELTA11	0.1127273E-03

As can be seen from the results, the solution to the linear program yields the following values for L_1, L_2, L_3, L_4, and L_5: 5.0000, 2.0800, 0.1127, 0.0000, and 0.0000, respectively. The optimum value of the objective function is 9.13×10^{-3}/s (approximately 288×10^3/yr). It is worth pointing out that the same optimum value of the objective function can also be achieved by other combinations of L_1 and L_2 along with the same value of L_3 (since both L_1 and L_2 are virtually free). The solution of P6.2 also yields a vanishing δ_4, indicating that the mass-exchange pinch is located at the line separating intervals 4 and 5. All these findings are consistent with the solutions obtained in Chapters Three and Five.

6.4 Optimization of Outlet Compositions

As has been discussed in Chapter Three, the target compositions are only upper bounds on the outlet compositions. Therefore, it may be necessary to optimize the outlet compositions.[1] A short-cut method of optimizing the outlet composition is the use of "lean substreams" (El-Halwagi, 1993; Garrison *et al.*, 1995). Consider an MSA, j, whose target composition is given by x_j^t. In order to determine the optimal outlet composition, x_j^{out}, a number, ND_j, of substreams are assumed. Each substream, d_j, *where* $d_j = 1, 2, \ldots, ND_j$, is a decomposed portion of the MSA which extends from the given x_j^s to a selected value of outlet composition,

---

[1] More rigorous techniques for optimizing outlet composition are described by El-Halwagi and Manousiouthakis (1990b), Garrison *et al.* (1995), and Gupta and Manousiouthakis (1995) and is beyond the scope of this book.

x^{out}_{j,d_j}, which lies between x^s_j and x^t_j. The flowrate of each substream, L_{j,d_j}, is unknown and is to be determined as part of the optimization problem. The number of substreams is dependent on the level of accuracy needed for the MEN analysis. Theoretically, an infinite number of substreams should be used to cover the whole composition span of each MSA. However, in practice few (typically less than five) substreams are needed. On the CID, the various substreams are represented against their composition scale. The formulation (P6.1) can, therefore, be revised to:

$$min \sum_{j=1}^{N_S} C_j \sum_{d_j=1}^{ND_j} L_{j,d_j} \qquad \text{(P6.3)}$$

subject to

$$\delta_k - \delta_{k-1} + \sum_{\substack{j \text{ passes through} \\ \text{interval } k}} \sum_{d_j=1}^{ND_j} L_{j,d_j} w^S_{j,k} = W^R_k \quad k = 1, 2, \ldots, N_{int} \quad L_{j,d_j} \geq 0,$$

$$j = 1, 2, \ldots, N_S$$

$$\sum_{d_j=1}^{ND_j} L_{j,d_j} \leq L^C_j, \quad j = 1, 2, \ldots, N_S$$

$$\delta_0 = 0$$

$$\delta_{N_{int}} = 0$$

$$\delta_k \geq 0, \quad k = 1, 2, \ldots, N_{int} - 1.$$

The above program (P6.3) is a linear program which minimizes the operating cost of MSAs. The solution of this program determines the optimal flowrate of each substream and, consequently, the optimal outlet compositions. If more than one substream are selected, the total flowrate can be obtained by summing up the individual flowrates of the substreams while the outlet composition may be determined by averaging the outlet compositions as follows:

$$L_j = \sum_{d_j=1}^{ND_j} L_{j,d_j}, \qquad \text{(6.7)}$$

$$x^{out}_j = \frac{\sum_{d_j=1}^{ND_j} L_{j,d_j} x^{out}_{j,d_j}}{L_j}. \qquad \text{(6.8)}$$

In order to demonstrate this procedure, let us revisit Example 3.1 on the recovery of benzene from a gaseous emission of a polymer facility. Instead of

Table 6.5
Data for Waste Stream in Benzene Removal Example

Stream	Flowrate G_i (kg mol/s)	Supply composition (mole fraction) y_i^s	Target composition (mole fraction) y_i^t
R_1	0.2	0.0020	0.0001

determining the outlet composition of S_1 graphically, it will be determined mathematically. The stream data for the waste stream and for the lean streams are given in Tables 6.5 and 6.6.

In order to determine the optimal outlet composition of S_1, several substreams are created to span an outlet composition between x_1^s and x_1^t. Let us select six substreams with outlet compositions of 0.0060, 0.0055, 0.0050, 0.0045, 0.0040, and 0.0035. The CID for the problem is shown by Fig. 6.2.

In terms of the LINGO input, the problem can be formulated via the following linear program:

```
MODEL:
MIN = 0.05*L3;

D1 + 0.0005*L2 = 5.0E-05;
D2 - D1 + 0.0005*L11 + 0.00025*L2 = 2.5E-05;
D3 - D2 + 0.0005*L11 + 0.00025*L2 + 0.0005*L12
= 2.5E-05;
```

Table 6.6
Data for Lean Streams in Benzene Removal Example

Stream	Upper bound on flowrate L_j^C (kg mol/s)	Supply composition of benzene (mole fraction) x_j^s	Target composition of benzene (mole fraction) x_j^t	m_j	C_j ($/kmol)	ε_j
S_1	0.08	0.0030	0.0060	0.25	0.00	0.0010
S_2	0.05	0.0020	0.0040^a	0.50	0.00	0.0010
S_3	∞	0.0008	0.0085	0.10	0.05	0.0002

[a]This value is located above the inlet of R_1 on the CID and therefore must be reduced. Since R_1 has a supply composition of 0.002, the maximum practically feasible value of x_2^t is $(0.002/0.5) - 0.001 = 0.003$.

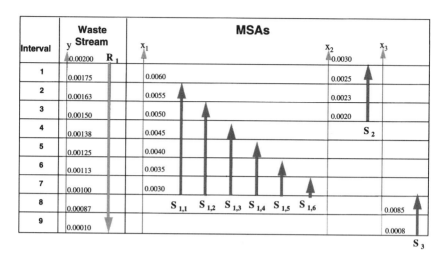

Figure 6.2 CID for benzene recovery example with lean substreams.

D4 - D3 + 0.0005*L11 + 0.0005*L12 + 0.0005*L13
= 2.5E-05;
D5 - D4 + 0.0005*L11 + 0.0005*L12 + 0.0005*L13
+ 0.0005*L14 = 2.5E-05;
D6 - D5 + 0.0005*L11 + 0.0005*L12 + 0.0005*L13
+ 0.0005*L14 + 0.0005*L15 = 2.5E05;
D7 - D6 + 0.0005*L11 + 0.0005*L12 + 0.0005*L13
+ 0.0005*L14 + 0.0005*L15 + 0.0005*L16 = 2.5E-05;
D8 - D7 = 2.6E-05;
- D8 + 0.0077*L3 = 0.000154;
D1 > 0.0;
D2 > 0.0;
D3 > 0.0;
D4 > 0.0;
D5 > 0.0;
D6 > 0.0;
D7 > 0.0;
D8 > 0.0;
L11 > 0.0;
L12 > 0.0;
L13 > 0.0;
L14 > 0.0;
L15 > 0.0;
L16 > 0.0;

```
L2 > 0.0;
L2 < 0.05;
L3 > 0.0;
L11 + L12 + L13 + L14 + L15 + L16 < 0.08;
END
```

The solution to this program yields the following results:

```
Objective value:      0.1168831E-02
```

Variable	Value
L3	0.2337662E-01
D1	0.5000000E-04
L2	0.0000000E+00
D2	0.4300000E-04
L11	0.6400000E-01
D3	0.3600000E-04
L12	0.0000000E+00
D4	0.2900000E-04
L13	0.0000000E+00
D5	0.2200000E-04
L14	0.0000000E+00
D6	0.1500000E-04
L15	0.0000000E+00
D7	0.0000000E+00
L16	0.1600000E-01
D8	0.2600000E-04

This solution indicates that the minimum operating cost is \$33,700/yr (\$ 0.00117/s) which corresponds to an optimal flowrate of S_3 of 0.0234 kg mol/s. By applying Eqs. (6.7) and (6.8), we can determine the flowrate of S_1 to be 0.08 kg mol/s and the outlet composition to be 0.0055. The pinch location corresponds to the vanishing residual mass at the line separating intervals seven and eight ($y = 0.001$). All these results are consistent with those obtained graphically in Chapter Three. Once again, more than one solution may be obtained to give the same value of the objective function.

6.5 Stream Matching and Network Synthesis

Having identified the values of all the flowrates of lean streams as well as the pinch location, we can now minimize the number of mass exchangers for a MOC solution. As has been previously mentioned, when a pinch point exists, the synthesis problem can be decomposed into two subnetworks, one above the pinch and one below the pinch. The subnetworks will be denoted by SN_m, where $m = 1,2$. It is, therefore, useful to define the following subsets:

$$R_m = \{i \mid i \in R, \text{stream } i \text{ exists in } SN_m\} \tag{6.9}$$

$$S_m = \{j \mid j \in S, \text{stream } j \text{ exists in } SN_m\} \tag{6.10}$$

$$R_{m,k} = \{i \mid i \in R_m, \text{stream } i \text{ exists in interval } \bar{k} \le k; \bar{k}, k \in SN_m\} \tag{6.11}$$

$$S_{m,k} = \{j \mid j \in 0S_m, \text{stream } j \text{ exists in interval } k \in SN_m\}. \tag{6.12}$$

For a rich stream, i,

$$\delta_{i,k} - \delta_{i,k-1} + \sum_{j \in S_{m,k}} W_{i,j,k} = W_{i,k}^R.$$

Within any subnetwork, the mass exchanged between any two streams is bounded by the smaller of the two loads. Therefore, the upper bound on the exchangeable mass between streams i and j in SN_m is given by

$$U_{i,j,m} = \min \left\{ \sum_{k \in SN_m} W_{i,k}^R, \sum_{k \in SN_m} W_{j,k}^S \right\}. \tag{6.13}$$

Now, we define the binary variable $E_{i,j,m}$, which takes the values of 0 when there is no match between streams i and j in SN_m, and takes the value of 1 when there exists a match between streams i and j (and hence an exchanger) in SN_m. Based on Eq. (6.13), one can write

$$\sum_{k \in SN_m} W_{i,j,k} - U_{i,j,m} \le 0 \quad i \in R_m, \ j \in S_m, \ m = 1, 2, \tag{6.14}$$

where $W_{i,j,k}$ denotes the mass exchanged between the ith rich stream and the jth lean stream in the kth interval. Therefore, the problem of minimizing the number of mass exchangers can be formulated as a mixed integer linear program "MILP" (El-Halwagi and Manousiouthakis, 1990a):

$$\text{minimize} \sum_{m=1,2} \sum_{i \in R_m} \sum_{j \in S_m} E_{i,j,m}, \tag{P6.4}$$

subject to the following:

Material balance for each rich stream around composition intervals:

$$\delta_{i,k} - \delta_{i,k-1} + \sum_{j \in S_{m,k}} W_{i,j,k} = W_{i,k}^R \quad i \in R_{m,k}, \ k \in SN_m, \ m = 1, 2$$

Material balance for each lean stream around composition intervals:

$$\sum_{i \in R_{m,k}} W_{i,j,k} = W_{j,k}^S \quad j \in S_{m,k}, \ k \in SN_m, \ m = 1, 2$$

Matching of loads:

$$\sum_{k \in SN_m} W_{i,j,k} - U_{i,j,m}E_{i,j,m} \leq 0 \quad i \in R_m, \, j \in S_m, \, m = 1, 2$$

Non-negative residuals

$$\delta_{i,k} \geq 0 \quad i \in R_{m,k}, \, k \in SN_m, \, m = 1, 2$$

Non-negative loads:

$$W_{i,j,k} \geq 0 \quad i \in R_{m,k}, j \in S_{m,k}, k \in SN_m, \, m = 1, 2$$

Binary integer variables for matching streams:

$$E_{i,j,m} = 0/1 \quad i \in R_m, j \in S_m, \, m = 1, 2$$

The above program is an MILP that can be solved (e.g., using the computer code LINGO) to provide information on the stream matches and exchangeable loads. It is interesting to note that the solution of program (P6.4) may not be unique. It is possible to generate all integer solutions to P6.4 by adding constraints that exclude previously obtained solutions from further consideration. For example, any previous solution can be eliminated by requiring that the sum of $E_{i,j,m}$ that were nonzero in that solution be less than the minimum number of exchangers. It is also worth mentioning that if the costs of the various exchangers are significantly different, the objective function can be modified by multiplying each integer variable by a weighing factor that reflects the relative cost of each unit.

6.6 Network Synthesis for Dephenolization Example

Let us revisit the dephenolization problem described in Sections 3.2 and 6.3. The objective is to synthesize a MOC-MEN with the least number of units. First, CID (Fig. 6.3) and the tables of exchangeable loads "TEL" (Tables 6.7 and 6.8) are developed based on the MOC solution identified in Sections 3.2 and 6.3. Since neither S_4 nor S_5 were selected as part of the MOC solution, there is no need to include them. Furthermore, since the optimal flowrates of S_1, S_2 and S_3 have been determined, the TEL for the MSAs can now be developed with the total loads of MSAs and not per kg of each MSA.

Since the pinch decomposes the problem into two subnetworks; it is useful to calculate the exchangeable load of each stream above and below the pinch. These values are presented in Tables 6.9 and 6.10.

We can now formulate the synthesis task as an MILP whose objective is to minimize the number of exchangers. Above the pinch (subnetwork $m = 1$),

Table 6.7
TEL for Waste Streams in Dephenolization Example

	Interval	Load of waste streams (kg phenol/s)	
		R_1	R_2
	1	0.0052	–
	2	0.0308	–
	3	0.0040	–
	4	0.0264	0.0132
Pinch			
	5	0.0096	0.0048
	6	0.0040	0.0020
	7	–	0.0040
	8	–	–
	9	–	–

there are four possible matches: R_1-S_1, R_1-S_2, R_2-S_1 and R_2-S_2. Hence, we need to define four binary variables ($E_{1,1,1}$, $E_{1,2,1}$, $E_{2,1,1}$, and $E_{2,2,1}$). Similarly, below the pinch (subnetwork $m = 2$) we have to define four binary variables ($E_{1,1,2} + E_{1,3,2} + E_{2,1,2} + E_{2,3,2}$) to represent potential matches between R_1-S_1,

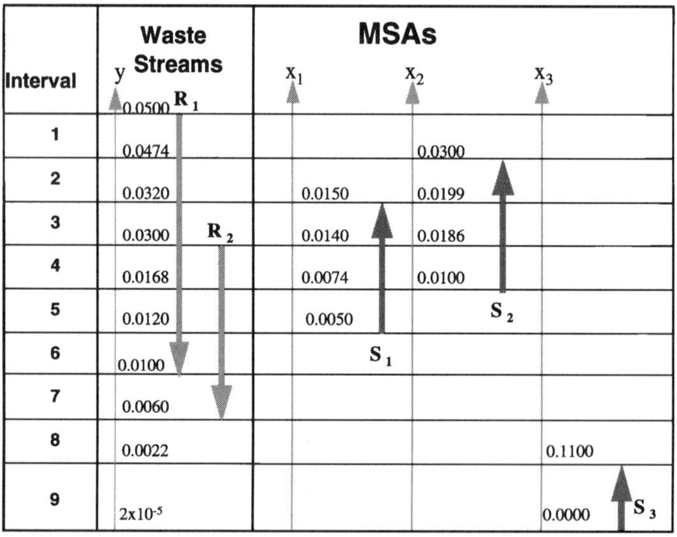

Figure 6.3 The CID for the dephenolization problem.

Table 6.8
TEL for MSAs in Dephenolization Example

	Interval	Load of MSAs (kg phenol/s)		
		S_1	S_2	S_3
	1	–	–	–
	2	–	0.0210	–
	3	0.0050	0.0027	–
	4	0.0330	0.0179	–
Pinch				
	5	0.0120	–	–
	6	–	–	–
	7	–	–	–
	8	–	–	–
	9	–	–	0.0124

Table 6.9
Exchangeable Loads Above the Pinch

Stream	Load (kg Phenol/s)
R_1	0.0664
R_2	0.0132
S_1	0.0380
S_2	0.0416
S_3	0.0000

Table 6.10
Exchangeable Loads Below the Pinch

Stream	Load (kg Phenol/s)
R_1	0.0136
R_2	0.0108
S_1	0.0120
S_2	0.0000
S_3	0.0124

R_1-S_3, R_2-S_1 and R_2-S_3. Therefore, the objective function is described by

Minimize $E_{1,1,1} + E_{1,2,1} + E_{2,1,1} + E_{2,2,1} + E_{1,1,2} + E_{1,3,2} + E_{2,1,2} + E_{2,3,2}$

subject to the following constraints:

Material balances for R_1 around composition intervals:

$$\delta_{1,1} = 0.0052$$
$$\delta_{1,2} - \delta_{1,1} + W_{1,2,2} = 0.0308$$
$$\delta_{1,3} - \delta_{1,2} + W_{1,1,3} + W_{1,2,3} = 0.0040$$
$$\delta_{1,4} - \delta_{1,3} + W_{1,1,4} + W_{1,2,4} = 0.0264$$
$$\delta_{1,5} - \delta_{1,4} + W_{1,1,5} = 0.0096$$
$$\delta_{1,6} - \delta_{1,5} = 0.0040$$
$$\delta_{1,7} - \delta_{1,6} = 0.0000$$
$$\delta_{1,8} - \delta_{1,7} = 0.0000$$
$$-\delta_{1,8} + W_{1,3,9} = 0.0000$$

Material balances for R_2 around composition intervals:

$$\delta_{2,4} + W_{2,1,4} + W_{2,2,4} = 0.0132$$
$$\delta_{2,5} - \delta_{2,4} + W_{2,1,5} = 0.0048$$
$$\delta_{2,6} - \delta_{2,5} = 0.0020$$
$$\delta_{2,7} - \delta_{2,6} = 0.0040$$
$$\delta_{2,8} - \delta_{2,7} = 0.0000$$
$$-\delta_{2,8} + W_{2,3,9} = 0.0000$$

Material balances for S_1 around composition intervals:

$$W_{1,1,3} = 0.0050$$
$$W_{1,1,4} + W_{2,1,4} = 0.0330$$
$$W_{1,1,5} + W_{2,1,5} = 0.0120$$

Material balances for S_2 around composition intervals:

$$W_{1,2,2} = 0.0210$$
$$W_{1,2,3} = 0.0027$$
$$W_{1,2,4} + W_{2,2,4} = 0.0179$$

Material balances for S_3 around the ninth interval:

$$W_{1,3,9} + W_{2,3,9} = 0.0124$$

Matching of loads:

$$W_{1,1,3} + W_{1,1,4} \leq 0.0380E_{1,1,1}$$

$$W_{1,2,2} + W_{1,2,3} + W_{1,2,4} \leq 0.0416E_{1,2,1}$$

$$W_{2,1,4} \leq 0.0132E_{2,1,1}$$

$$W_{2,2,4} \leq 0.0132E_{2,2,1}$$

$$W_{1,1,5} \leq 0.0120E_{1,1,2}$$

$$W_{2,1,5} \leq 0.0108E_{2,1,2}$$

$$W_{1,3,9} \leq 0.0124E_{1,3,2}$$

$$W_{2,3,9} \leq 0.0108E_{2,3,2}$$

with the non-negativity and integer constraints.

In terms of LINGO input, the above program can be written as follows:

```
MODEL:
MIN = E111 + E121 + E211 + E221 + E112 + E132 + E212
    + E232;

D11 = 0.0052;
D12 - D11 + W122 = 0.0308;
D13 - D12 + W113 + W123 = 0.0040;
D14 - D13 + W114 + W124 = 0.0264;
D15 - D14 + W115 = 0.0096;
D16 - D15 = 0.0040;
D17 - D16 = 0.0000;
D18 - D17 = 0.0000;
  - D18 + W139 = 0.0000;
D24 - W214 + W224 = 0.0132;
D25 - D24 + W215 = 0.0048;
D26 - D25 = 0.0020;
D27 - D26 = 0.0040;
D28 - D27 = 0.0000;
  - D28 + W239 = 0.0000;
W113 = 0.0050;
W114 + W214 = 0.0330;
W115 + W215 = 0.0120;
W122 = 0.0210;
```

```
W123 = 0.0027;
W124 + W224 = 0.0179;
W139 + W239 = 0.0124;
W113 + W114 <= 0.038*E111;
W122 + W123 + W124 <= 0.0416*E121;
W214 <= 0.0132*E211;
W224 <= 0.0132*E221;
W115 <= 0.012*E112;
W215 <= 0.0108*E212;
W139 <= 0.0124*E132;
W239 <= 0.0108*E232;
D11 >= 0.0;
D12 >= 0.0;
D13 >= 0.0;
D14 >= 0.0;
D15 >= 0.0;
D16 >= 0.0;
D17 >= 0.0;
D18 >= 0.0;
D24 >= 0.0;
D25 >= 0.0;
D26 >= 0.0;
D27 >= 0.0;
D28 >= 0.0;
W122 >= 0.0;
W113 >= 0.0;
W123 >= 0.0;
W124 >= 0.0;
W139 >= 0.0;
W214 >= 0.0;
W224 >= 0.0;
W215 >= 0.0;
W239 >= 0.0;
@BIN(E111);
@BIN(E121);
@BIN(E211);
@BIN(E221);
@BIN(E112);
@BIN(E132);
@BIN(E212);
@BIN(E232);
END
```

This MILP can be solved using LINGO to yield the following results:

```
Optimal solution found at step:    13
Objective value:                   7.000000
Branch count:                      0
```

Variable	Value
E111	1.000000
E121	1.000000
E211	0.000000
E221	1.000000
E112	1.000000
E132	1.000000
E212	1.000000
E232	1.000000
D11	0.5200000E-02
D12	0.1500000E-01
W122	0.2100000E-01
D13	0.1130000E-01
W113	0.5000000E-02
W123	0.2700000E-02
D14	0.0000000E+00
W114	0.3300000E-01
W124	0.4699999E-02
D15	0.2400000E-02
W115	0.7200001E-02
D16	0.6400000E-02
D17	0.6400000E-02
D18	0.6400000E-02
W139	0.6400000E-02
D24	0.0000000E+00
W214	0.00000000E+00
W224	0.1320000E-01
D25	0.0000000E+00
W215	0.4800000E-02
D26	0.2000000E-02
D27	0.6000000E-02
D28	0.6000000E-02
W239	0.6000000E-02

These results indicate that the solution features seven units that represent matches between R_1-S_1, R_1-S_2, and R_2-S_2 above the pinch and R_1-S_1, R_1-S_3, R_2-S_1 and R_2-S_3 below the pinch. The load for each exchanger can be evaluated by simply adding up the exchangeable loads within the same subnetwork. For

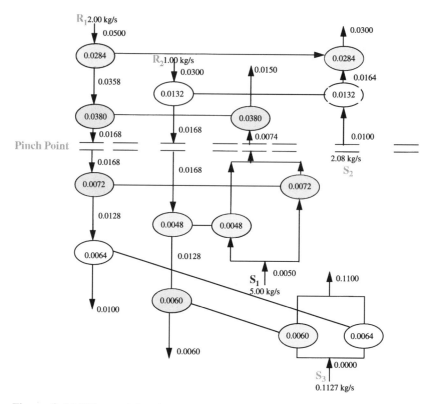

Figure 6.4 MOC network for dephenolization example.

instance, the transferable load from R_1 to S_2 above can be calculated as follows:

$$\text{Exchangeable load for } E_{1,2,1} = W_{1,2,2} + W_{1,2,3} + W_{1,2,4}$$
$$= 0.0210 + 0.0027 + 0.0047$$
$$= 0.0284 \text{ kg phenol/s} \qquad (6.15)$$

Hence, one can use these results to construct the network shown in Fig. 6.4. This is the same configuration obtained using the algebraic method as illustrated by Fig. 5.13. However, the loads below the pinch are distributed differently. This is consistent with the previously mentioned observation that multiple solutions featuring same objective function can exist for the problem. The final design should be based on considerations of total annualized cost, safety, flexibility, operability and controllability. As has been discussed in Sections 3.11 and 5.8, the minimum TAC can be attained by trading off fixed versus operating costs by optimizing driving forces, stream mixing and mass-load paths.

Problems

6.1 Using linear programming, resolve the dephenolization example presented in this chapter for the case when the two waste streams are allowed to mix.

6.2 Using mixed-integer programming, find the minimum number of mass exchangers the benzene recovery example described in Section 3.7 (Example 3.1).

6.3 Employ linear programming to find the MOC solution for the toluene-removal example described in Section 3.10 (Example 3.3).

6.4 Use optimization to solve problem 3.1.

6.5 Apply the optimization-based approach presented in this chapter to solve problem 3.3.

6.6 Employ linear programming to solve problem 3.4.

6.7 Solve problem 5.5 using optimization.

6.8 Consider the metal pickling plant shown in Fig. 6.5 (El-Halwagi and Manousiouthakis, 1990a). The objective of this process is to use a pickle solution (e.g., HCl) to remove corrosion products, oxides and scales from the metal surface. The spent pickle solution

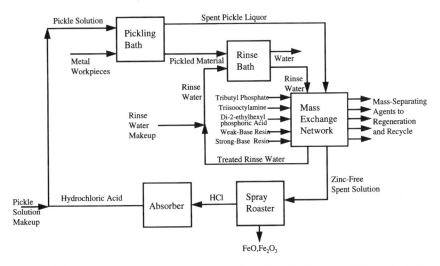

Figure 6.5 Zinc recovery from metal picking plant from (from El-Halwagi and Manousiouthakis, 1990a. Automatic Synthesis of MassExchange Networks, *Chem. Eng. Sci.*, 45(9), p. 2818, Copyright © 1990, with kind permission from Elsevier Science Ltd., The Boulevard, Langford Lane, Kidlington OX5 1GB, UK.)

Table 6.11
Stream Data for the Zinc Recovery Problem

	Rich stream				MSAs				
Stream	G_i (kg/s)	y_i^s	y_i^t	Stream	L_j^c (kg/s)	x_j^s	x_j^t	m_j	b_j	c_j ($/kg)
R_1	0.2	0.08	0.02	S_1		0.0060	0.0600	0.845	0.000	0.02
R_2	0.1	0.03	0.001	S_2		0.0100	0.0200	1.134	0.010	0.11
				S_3		0.0090	0.0500	0.632	0.020	0.04
				S_4		0.0001	0.0100	0.376	0.0001	0.05
				S_5		0.0040	0.0150	0.362	0.002	0.13

contains zinc chloride and ferrous chloride as the two primary contaminants. Mass exchange can be used to selectively recover zinc chloride from the spent liquor. The zinc-free liquor is then forwarded to a spray furnace in which ferrous chloride is converted to hydrogen chloride and iron oxides. The hydrogen chloride is absorbed and recycled to the pickling path. The metal leaving the pickling path is rinsed off by water to remove the clinging film of drag-out chemicals that adheres to the workpiece surface. The rinse wastewater contains zinc chloride as the primary pollutant that must be recovered for environmental and economic purposes.

The purpose of the problem is to systematically synthesize a cost-effective MEN that can recover zinc chloride from the spent pickle liquor, R_1, and the rinse wastewater, R_2. Two mass-exchange processes are proposed for recovering zinc; solvent extraction and ion exchange. For solvent extraction, three candidate MSAs are suggested: tributyl phosphate, S_1, triisooctyl amine, S_2, and di-2-ethyl hexyl phosphoric acid, S_3. For ion exchange, two resins are proposed: a strong-base resin, S_4, and a weak-base resin, S_5. Table 6.11 summarizes the data for all the streams. All compositions are given in mass fractions. Assume a value of 0.0001 for the minimum allowable composition difference for all lean streams.

6.9 Etching of copper, using an ammoniacal solution, is an important operation in the manufacture of printed circuit boards for the microelectronics industry. During etching, the concentration of copper in the ammoniacal solution increases. Etching is most efficiently carried out for copper concentrations between 10 and 13 w/w% in the solution while etching efficiency almost vanishes at higher concentrations (15–17 w/w%). In order to maintain the etching efficiency, copper must be continuously removed from the spent ammoniacal solution through solvent extraction. The regenerated ammoniacal etchant can then be recycled to the etching line.

The etched printed circuit boards are washed out with water to dilute the concentration of the contaminants on the board surface to an acceptable level. The extraction of copper from the effluent rinse water is essential for both environmental and economic reasons since decontaminated water is returned to the rinse vessel.

A schematic representation of the etching process is demonstrated in Fig. 6.6. The proposed copper recovery scheme is to feed both the spent etchant and the effluent rinse water

Table 6.12
Stream Data for the Copper Etching Problem

	Rich streams				MSAs				
Stream	G_i (kg/s)	y_i^s	y_i^t	Stream	L_j^c (kg/s)	x_j^s	x_j^t	m_j	b_j	C_j ($/kg)
R_1	0.25	0.13	0.10	S_1	∞	0.030	0.070	0.734	0.001	0.01
R_2	0.10	0.06	0.02	S_2	∞	0.001	0.020	0.111	1.013	0.12

to a MEN in which copper is transferred to some selective solvents. Two extractants are recommended for this separation task: LIX63 (an aliphatic-hydroxyoxime), S_1, and P_1 (an aromatic-hydroxyoxime), S_2. The former solvent appears to work most efficiently at moderate copper concentrations, whereas the latter extractant offers remarkable extraction efficiencies at low copper concentrations. Table 6.12 summarizes the stream data for the problem.

Two types of contractors will be utilized: a perforated-plate column for S_1 and a packed column for S_2. The basic design and cost data that should be employed in this problem are given by El-Halwagi and Manousiouthakis (Chem. Eng. Sci., 45(9), p. 2831, 1990a).

It is desired to synthesizes an optimum MEN that features minimum total annualized cost, "TAC", where

$$TAC = \text{Annualized fixed cost} + \text{Annual operating cost}.$$

(Hint: vary the minimum allowable composition differences to iteratively trade off fixed versus operating costs).

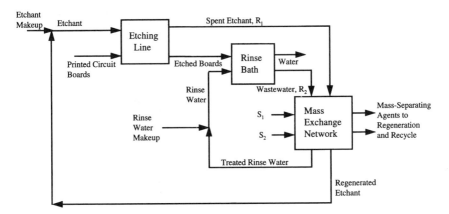

Figure 6.6 Recovery of Copper from Liquid Effluents of an Etching Plant (El-Halwagi and Manousiouthakis, 1990a. Automatic Synthesis of MassExchange Networks, *Chem. Eng. Sci.*, 45(9), p. 2825, Copyright © 1990, with kind permission from Elsevier Science Ltd., The Boulevard, Langford Lane, Kidlington 0X5 1GB, UK.)

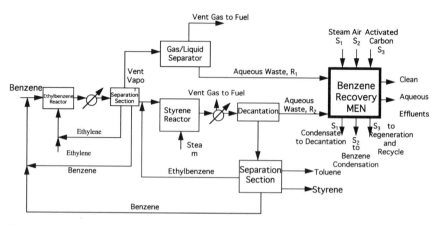

Figure 6.7 Schematic flowsheet of a styrene plant (Stanley and El-Halwagi, 1995, reproduced with permission of the McGraw Hill Companies).

6.10 Styrene can be produced by the dehydrogenation of ethylbenzene using live steam over an oxide catalyst at temperatures of 600°C to 650°C (Stanley and El-Halwagi, 1995). The process flowsheet is shown by Fig. 6.7. The first step in the process is to convert ethylene and benzene into ethylbenzene. The reaction products are cooled and separated. One of the separated streams is an aqueous waste (R_1). The main pollutant in this stream is benzene. Ethylbenzene leaving the separation section is fed to the styrene reactor whereby styrene and hydrogen are formed. Furthermore, by products (benzene, ethane, toluene and methane) are generated. The reactor product is then cooled and decanted. The aqueous layer leaving the decanter is a wastewater stream (R_2) which consists of steam condensate saturated with benzene. The organic layer, consisting of styrene, benzene, toluene, and unreacted ethylbenzene, is sent to a separation section.

There are two primary sources for aqueous pollution in this process—the condensate streams R_1 (1,000 kg/hr) and R_2 (69,500 kg/hr). Both streams have the same supply composition, which corresponds to the solubility of benzene in water which is 1770 ppm (1.77×10^{-3} kg benzene/kg water). Consequently, they may be combined as a single stream. The target composition is 57 ppb as dictated by the VOC environmental regulations called NPDES (National Pollutant Discharge Elimination System).

Three mass-exchange operations are considered: steam stripping, air stripping and adsorption using granular activated carbon. The stream data are given in Table 6.13.

Using linear programming, determine the MOC solution of the system.

6.11 In the previous problem, it is desired to compare the total annualized cost of the benzene-recovery system to the value of recovered benzene. The total annualized cost "TAC" for the network is defined as:

$$\text{TAC} = \text{Annual operating cost} + 0.2 \times \text{Fixed capital cost}.$$

The fixed cost($) of a moving-bed adsorption or regeneration column is given by $30,000V^{0.57}$, where V is the volume of the column (m³) based on a 15-minute residence

Table 6.13
Data of the MSAs in Styrene Plant Problem

Stream	L_j^C (kg/s)	Supply composition x_j^s	Target composition x_j^t	m_j	ε_j	C_j ($/kg MSA)
Stream (S_1)	∞	0	1.62	0.5	0.15	0.004
Air (S_2)	∞	0	0.02	1.0	0.01	0.003
Carbon(S_3)	∞	3×10^{-5}	0.20	0.8	5×10^{-5}	0.026

All compositions and equilibrium data are in **mass ratios,** kg benzene/kg benzene-free MSA.

time for the combined flowrate of carbon and wastewater (or steam). A steam stripper already exists on site with its piping, ancillary equipment and instrumentation. The column will be salvaged for benzene recovery. The only changes needed involve replacing the plates inside the column with new sieve trays. The fixed cost of the sieve trays is $1,750/plate. The overall column efficiency is assumed to be 65%.

If the value of recovered benzene is taken as $0.20/kg, compare the annual revenue from recovering benzene to the TAC of the MEN.

6.12 In many situations, there is a trade-off between reducing the amount of generated waste at the source versus its recovery via a separation system. For instance, in problem 6.10, live steam is used in the styrene reactor to enhance the product yield. However, the steam eventually constitutes the aqueous waste, R_2. Hence, the higher the flowrate of steam, the larger the cost of the benzene recycle/reuse MEN. These opposing effects call for the simultaneous consideration of source reduction of R_2 along with its recycle/reuse. One way of approaching this problem is by invoking economic criteria. Let us define the economic potential of the process ($/yr) as follows (Stanley and El-Halwagi, 1995):

Economic potential = Value of produced styrene
+ Value of recovered ethylbenzene
− Cost of Ethylbenzene − Cost of Steam
− TAC of the recycle/reuse network

Determine the optimal steam ratio (kg steam/kg ethylbenzene) that should be used in the styrene reactor in order to maximize the economic potential of the process.

Symbols

C_j unit cost of the jth MSA including regeneration and makeup, $/kg of recirculating MSA)

d_j index for substreams of the jth MSA

$E_{i,j,m}$ a binary integer variable designating the existence or absence of an exchanger between rich stream i and lean stream j in subnetwork m

G_i flowrate of the ith waste stream

i index of waste streams

j index of MSAs

k index of composition intervals

L_j flowrate of the jth MSA(kg/s)

L_j, d_j flowrate of substream of d_j the jth MSA(kg/s)

L_{j^c} upper bound on available flowrate of the jth MSA(kg/s)

m subnetwork (one above the pinch and two below the pinch)

m_j slope of equilibrium line for the jth MSA

N_{int} number of composition intervals

N_R number of waste streams

N_S number of MSAs

ND_j Number of substreams for the jth MSA

R_i the ith waste stream

R_m a set defined by Eq. 6.9

$R_{m,k}$ a set defined by Eq. 6.11

S_j the jth MSA

S_m a set defined by Eq. 6.10

$S_{m,k}$ a set defined by Eq. 6.12

SN_m subnetwork m

$U_{i,j,m}$ upper bound on exchangeable mass between i and j in subnetwork m (defined by Eq. 6.13)(kg/s)

$W_{i,j,k}$ exchangeable load between the ith waste stream and the jth MSA in the kth interval (kg/s)

$W_{i,k}^R$ exchangeable load of the ith waste stream which passes through the kth interval as defined by Eqs. (6.2) and (6.3) (kg/s)

$w_{j,k}^S$ exchangeable load of the jth MSA which passes through the kth interval as defined by Eqs. (6.5) and (6.6) (kg/s)

W_k^R the collective exchangeable load of the waste streams in interval k as defined by Eq. (6.4) (kg/s)

$x_{j,k-1}$ composition of key component in the jth MSA at the upper horizontal line defining the kth interval

$x_{j,k}$ composition of key component in the jth MSA at the lower horizontal line defining the kth interval

x_j^{out} outlet composition of the jth MSA (defined by Eq. 6.8)

x_j^s supply composition of the jth MSA

x_j^t target composition of the jth MSA

y_{k-1} composition of key component in the ith waste stream at the upper horizontal line defining the kth interval

y_k composition of key component in the ith waste stream at the lower horizontal line defining the kth interval

References

El-Halwagi, M. M. (1993). A process synthesis approach to the dilemma of simultaneous heat recovery, waste reduction and cost effectiveness. In "Proceedings of the Third Cairo International Conference on Renewable Energy Sources" (A. I. El-Sharkawy and R. H. Kummler, eds.) Vol. 2, 579–594.

El-Halwagi, M. M. (1995). Introduction to numerical optimization approaches to pollution prevention. In "Waste Minimization Through Process Design," (A. P. Rossiter, ed.) pp. 199–208, McGraw Hill, New York.

El-Halwagi, M. M. and Manousiouthakis, V. (1990a). Automatic synthesis of mass-exchange networks with single-component targets. *Chem. Eng. Sci.* **45**(9), 2813–2831.

El-Halwagi, M. M. and Manousiouthakis, V. (1990b). Simultaneous synthesis of mass exchange and regeneration networks. *AIChE J.* **36**(8), 1209–1219.

Garrison, G. W., Cooley, B. L., and El-Halwagi, M. M. (1995). Synthesis of mass exchange networks with multiple target mass separating agents. *Dev. Chem. Eng. Min. Proc.* **3**(1), 31–49.

Gupta, A. and Manousiouthakis V. (1995). Mass-exchange networks with variable single component targets: minimum uutility cost through linear programming. *AIChE Annual Meeting*, Miami, November.

Stanley, C. and El-Halwagi M. M. (1995). Synthesis of mass-exchange networks using linear programming techniques. In "Waste Minimization Through Process Design" (A. P. Rossiter, ed.), pp. 209–225 McGraw Hill, New York.

CHAPTER SEVEN

Mathematical Optimization Techniques for Mass Integration

Chapter One presented an overview of mass integration strategies involving stream routing and species allocation, generation and interception. The graphical techniques described in Chapter Four provide the designer with a systematic framework for understanding the global insights of the process and employing them for cost-effective pollution prevention. This chapter presents mathematical optimization techniques for mass integration including mass-exchange interception and rerouting of targeted species throughout the process. Since the interception devices in this chapter are limited to mass-exchange operations, the waste-interception network (WIN) becomes a mass-exchange network (MEN). The emphasis is on targeting pollution at the heart of the plant rather than dealing with pollutants in terminal waste streams. The chapter also provides a mathematical framework for tackling pollution as a multimedia (multiphase) problem.

7.1 Problem Statement and Challenges

The problem addressed in this chapter can be briefly stated as follows:

Given a process with terminal gaseous and liquid wastes that contain a certain pollutant, it is desired to identify minimum-cost strategies for segregation, mixing, *in-plant* pollutant interception using mass-exchange operations, recycle, and sink/generator manipulation that can reduce the pollutant load and concentration in the terminal waste streams to a specified level (Fig. 7.1).

To identify the optimum solution for the above-mentioned problem, one should be able to answer the following challenging questions:

· Which phase(s) (gaseous, liquid) should be intercepted?

154

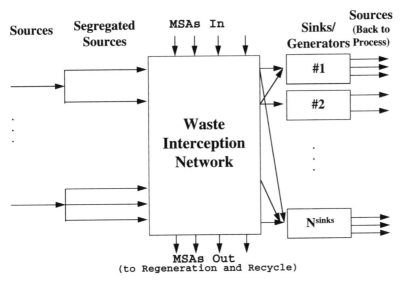

Figure 7.1 Schematic representation of mass-integration framework with MSA-induced interception (from El-Halwagi *et al.*, 1996. Reproduced with permission of the American Institute of Chemical Engineers. Copyright © 1996 AIChE. All rights reserved).

- Which process streams should be intercepted?
- To what extent should the pollutant be removed from each process stream?
- Which separation operations should be used for interception?
- Which MSAs should be selected for interception ?
- What is the optimal flowrate of each MSA?
- How should these separating agents be matched with the sources (i.e., stream pairings)?
- Which units (sinks/generators) should be manipulated for source reduction? By what means?
- Should any streams be segregated or mixed? Which ones?
- Which streams should be recycled/reused? To what units?

   Because of the complexity of these questions, the solution approach will be presented in stages. A systematic method for in-plant interception using MSAs is presented in the next section. Later, interception will be integrated with the other mass-integration strategies.

7.2 Synthesis of MSA-Induced WINs

In Chapters Three, Five, and Six, the MEN-synthesis techniques dealt with cases where the separation task was defined as part of the design task. Streams to be separated were given along with their supply and target compositions. Furthermore, separation of one source (rich stream) had no effect on the other sources. As has been discussed in Section 7.1., when mass integration involving in-plant interception is considered the designer has to determine which streams are to be intercepted and the extent of separation for each stream. Moreover, interception of one source may affect the other sources. Hence, the designer should understand the global flow of pollutants and mass interactions throughout the process and employ this understanding in setting optimal interception strategies. Towards this end, a particularly useful tool is the *path diagram*.

7.2.1 The Path Diagram

The path diagram (e.g. El-Halwagi *et al.*, 1996; El-Halwagi and Spriggs, 1996; Garrison *et al.*, 1995, 1996; Hamad *et al.*, 1995, 1996) is a representation tool that captures the overall flow of the pollutant throughout the plant. It also relates the flow of the pollutant to the performance of the different processing units. A path diagram represents the load of the targeted species throughout the process as a function of its composition in carrying streams. Instead of considering the whole flowsheet, one should keeps track of the units involving the targeted species. Hence, the pollutant is tracked throughout the process via material balances and unit modeling equations. A path diagram is created for each targeted species in each phase (gaseous, liquid, solid). Each pollutant-laden stream (source) is represented by a node on a load-composition diagram. These nodes are connected with composition profiles of the streams within the units. The exact shape of the composition profile within a unit is typically not needed unless modifications within the units are considered. Therefore, these profiles can be approximated by arrows of any shape. The directionality of these arrows reflects the orientation of mass flow: arrow tails emanate from inputs to units, and arrow heads point toward outputs of units. A unit may have multiple inputs and outputs, so each node may be associated with multiple arrow heads and tails.

In order to demonstrate the construction of the path diagram, consider Fig. 7.2 (El-Halwagi *et al.*, 1996), which illustrates a section of a process involving the pollutant laden streams. Figure 7.2 also shows the various loads and compositions of the pollutant throughout the process. For a given pollutant-laden stream v, the term V_v is the flowrate of the stream, y_v is the composition of the pollutant, and Φ_v is the load of the pollutant in the stream, defined as

$$\Phi_v = V_v y_v. \tag{7.1}$$

Similarly, on the liquid path, the load of the pollutant in the wth liquid source is given by

$$\Psi_w = W_w z_w, \tag{7.2}$$

where Ψ_w, W_w, and z_w are the load, flowrate, and composition of the wth liquid source (node).

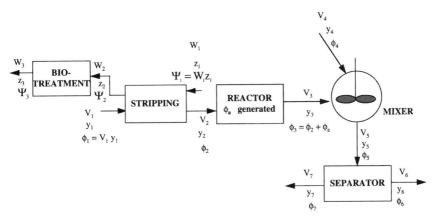

The path diagram for this process is given by Fig. 7.3. The first step in Fig. 7.3 is a stripping process in which the pollutant is transferred from a liquid stream (source $w = 1$) to a gaseous stream (source $v = 1$). Because of the countercurrent contact of the two streams, the path profile for each stream ($\Psi_w = W_w z_w$ and Φ_v versus y_v) is given by a linear arrow extending between inlet and outlet compositions and having a slope equal to the flowrate of the stream. The gaseous stream leaving the stripper is then processed in a continuously stirred tank reactor where

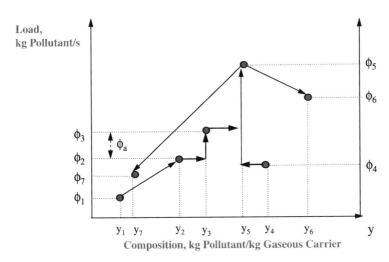

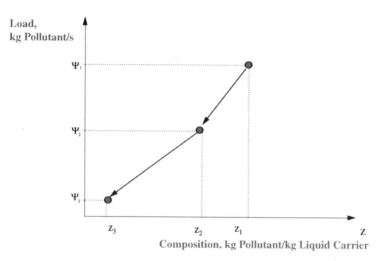

Figure 7.4 The liquid path diagram (from El-Halwagi *et al.*, 1996. Reproduced with permission of the American Institute of Chemical Engineers. Copyright © 1996 AIChE. All rights reserved).

some additional mass of the pollutant, Φ_a, is generated by chemical reaction. Owing to the complete mixing in the reactor, the concentration of the pollutant instantaneously changes from the inlet concentration, y_2, to the outlet concentration, y_3. Furthermore, the pollutant loading increases to $\Phi_3 = \Phi_2 + \Phi_a$. The effluent from the reactor is mixed with another process stream ($v = 4$) to give a composition y_5 and mass loading, Φ_5. The composition y_5 can be graphically determined using the lever-arm principle. The final operation involves the separation of the mixed stream into two terminal streams ($v = 6$ and 7).

Similarly, one can develop the liquid path diagram as shown in Fig. 7.4. It involves three nodes, $w = 1 - 3$, which are related by the stripper and the biotreatment units.

The path diagram provides the big picture for mass flow from a species viewpoint. This is a fundamentally different vision from the equipment-oriented description of a process (the flowsheet), in which the big picture is lost. The path diagram can also be used to determine the effect on the rest of the diagram of manipulating any node. In addition, as will be shown later, it provides a systematic way for identifying where to remove the pollutants and to what extent they should be removed.

7.2.2 Integration of the Path and the Pinch Diagrams

As mentioned before, the path diagram can be used to predict the effect of intercepting one stream on the rest of the streams. In order to quantify this relationship, let us consider that all gaseous and liquid nodes on the path diagrams are intercepted. Upon interception, the composition and load of the vth gaseous node are altered

Input Sources   MSAs in

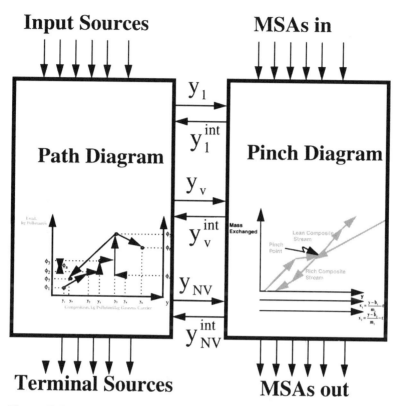

y_1

y_1^{int}

Path Diagram   Pinch Diagram

y_v

y_v^{int}

y_{NV}

y_{NV}^{int}

Terminal Sources   MSAs out

Figure 7.5 Integration of the path and the pinch diagrams (from El-Halwagi *et al.*, 1996. Reproduced with permission of the American Institute of Chemical Engineers. Copyright © 1996 AIChE. All rights reserved).

from y_v and Φ_v to y_v^{int} and Φ_v^{int}. Similarly, the composition and load of the wth liquid node are altered from z_w and Ψ_w to z_w^{int} and Ψ_w^{int}. The pinch diagram is used to determine optimal interceptions, whereas the path diagram is responsible for tracking the consequences of interceptions. The relationship between the path and the pinch diagrams can be envisioned as shown by Fig. 7.5 which illustrates the back-and-forth passage of streams between the two diagrams. Consider the gaseous nodes ($v = 1, 2, \ldots,$ NV) that are passed from the path diagram to the pinch diagram. Upon interception, these streams change composition from y_v to the intercepted compositions y_v^{int} where $v = 1, 2, \ldots,$ NV. Each intercepted stream is returned to the path diagram. According to the path-diagram equations, this interception propagates throughout the whole path diagram, affecting the other nodes. In turn, this propagation in the path diagram affects the pinch diagram by changing the waste composite stream. Consequently, the lean composite stream is adjusted to respond to the changes in the waste composite stream. The path and the pinch diagrams can be integrated using mixed-integer nonlinear programming (El-Halwagi *et al.*, 1996). Alternatively, one can use the shortcut method described in the following section.

7.2.3 Screening of Candidate MSAs Using a Hybrid of Path and Pinch Diagrams

The selection of interception technologies can be aided by useful concepts stemming from the path and the pinch diagrams. After the path diagram is plotted for the sources, the pinch diagram is superimposed through a horizontal projection. Composition scales for the MSAs are plotted in one-to-one correspondence with the composition scale for the sources using Eq. (3.5). Each MSA is then represented versus its composition scale as a horizontal arrow extending between its supply and target compositions (Fig. 7.6).

Several useful insights can be gained from this hybrid diagram. Let us consider two MSAs a and b, whose costs (\$/kg of recirculating MSA) are c_a and c_b. These costs can be converted into \$/kg of removed pollutant, c_a^r and c_b^r, as follows:

$$c_a^r = \frac{c_a}{x_a^t - x_a^s} \qquad (7.3a)$$

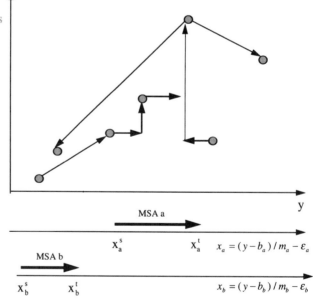

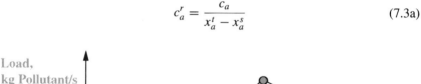

Figure 7.6 Screening MSAs using the path diagram (from El-Halwagi et al., 1996. Reproduced with permission of the American Institute of Chemical Engineers. Copyright © 1996 AIChE. All rights reserved).

and

$$c_b^r = \frac{c_b}{x_b^t - x_b^s}. \tag{7.3b}$$

If arrow b lies completely to the left of arrow a and c_b^r is less than c_a^r, MSA b is chosen, because it is thermodynamically and economically superior to MSA a. Another useful observation can be inferred from the relative location of the sources and the MSAs. If an MSA lies to the right of a source node, the MSA is not a candidate for intercepting this node, because mass exchange is infeasible.

Once optimal interception strategies are prescreened for each node, one can formulate the synthesis task as an optimization program whose objective is to minimize interception cost subject to the path-diagram equations and the prescreened interception strategies. The following example illustrates this approach.

7.3 Case Study: Interception of Chloroethanol in an Ethyl Chloride Process

Ethyl chloride (C_2H_5Cl) can be manufactured by catalytically reacting ethanol and hydrochloric acid (El-Halwagi *et al.*, 1996). Figure 7.7 is a simplified flowsheet of

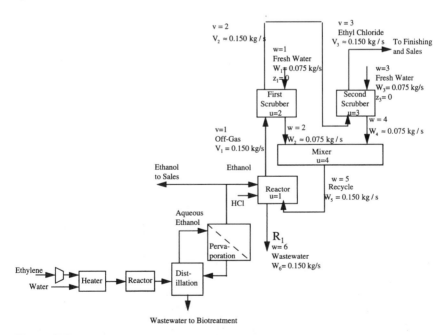

Figure 7.7 Flowsheet of the CE case study (from El-Halwagi et al., 1996. Reproduced with permission of the American Institute of Chemical Engineers. Copyright © 1996 AIChE. All rights reserved).

the process. The facility involves two integrated processes; the ethanol plant and the ethyl chloride plant. First, ethanol is manufactured by the catalytic hydration of ethylene. Compressed ethylene is heated with water and reacted to form ethanol. Ethanol is separated using distillation followed by membrane separation (pervaporation). The aqueous waste from distillation is fed to a biotreatment unit in which the organic content of the wastewater is used as a bionutrient. The separated ethanol is reacted with hydrochloric acid in a multiphase reactor to form ethyl chloride. The reaction takes place primarily in the liquid phase. A by-product of the reaction is chloroethanol (CE, also referred to as ethylene chlorohydrin, C_2H_5OCl). This side reaction will be referred to as the *oxychlorination reaction*. The rate of chloroethanol generation via oxychlorination (approximated by a pseudo zero-order reaction) is given

$$r_{oxychlorination} = 6.03 \times 10^{-6} \text{ kg chloroethanol/s.} \tag{7.4}$$

While ethyl chloride is one of the least toxic of all chlorinated hydrocarbons, CE is a toxic pollutant. The off-gas from the reactor is scrubbed with water in two absorption columns. The first column is intended to recover the majority of unreacted ethanol, hydrogen chloride, and CE. The second scrubber purifies the product from traces of unreacted materials and acts as a back-up column in case the first scrubber is out of operation. Each scrubber contains two sieve plates and has an overall column efficiency of 65% (i.e., NTP = 1.3). Following the scrubber, ethyl chloride is finished and sold. The aqueous streams leaving the scrubbers are mixed and recycled to the reactor. A fraction of the CE recycled to the reactor is reduced to ethyl chloride. This side reaction will be called the *reduction reaction*. The rate of CE depletion in the reactor due to this reaction can be approximated by the following pseudo first order expression:

$$r_{reduction} = 0.090 \ z_5 \text{ kg chloroethanol/s,} \tag{7.5}$$

where z_5 is in units of mass fraction.

The compositions of CE in the gaseous and liquid effluents of the ethyl chloride reactor are related through an equilibrium distribution coefficient as follows:

$$\frac{y_1}{z_6} = 5. \tag{7.6}$$

Because of the toxicity of CE, the aqueous effluent from the ethyl chloride reactor, R_1, causes significant problems for the bio-treatment facility. The objective of this case study is to optimally intercept CE-laden streams so as to reduce the CE content of R_1 to meet the following regulations:
Target composition:

$$z_6^{int} \leq 7 \ ppmw. \tag{7.7}$$

Table 7.1

Data for the MSAs That Can Remove CE from Gaseous Streams

Stream	Description	Supply composition x_j^s (ppmw)	Target composition x_j^t (ppmw)	m_j	ε_j ppmw	c_j $/kg MSA	c_j^r $/kg CE removed
SV_1	Polymeric resin	2	10	0.03	5	0.08	10,000
SV_2	Activated carbon	5	30	0.06	10	0.10	4,000
SV_3	Oil	200	300	0.80	20	0.05	500

Target load:

$$\text{Load of CE in } R_1 \leq 1.05 \times 10^{-6} \ kg \ CE/s. \tag{7.8}$$

Six MSAs are available for removing CE: three for gaseous streams and three for liquid streams (Tables 7.1 and 7.2).

Solution

This case study involves four units, three gaseous sources (nodes), and six liquid sources. The flowrates of all gaseous and liquid sources are shown in Fig. 7.7. Also, the compositions of the two liquid nodes corresponding to entering fresh water are given ($z_1 = z_3 = 0$). The first step in the analysis is the development of the path-diagram equations to quantify the relationship of the seven unknown compositions. As shown by Fig. 7.7, overall material balances around the four units ($u = 1-4$, assuming constant flowrate of carriers) provide the following flowrates of the carrier gas and liquid:

$$V_1 = V_2 = V_3 = 0.150 \ \text{kg/s} \tag{7.9}$$

$$W_1 = W_2 = W_3 = W_4 = 0.075 \ \text{kg/s} \tag{7.10}$$

$$W_5 = W_6 = 0.150 \ \text{kg/s}. \tag{7.11}$$

Table 7.2

Data for the MSAs That Can Remove CE from Liquid Streams

Stream	Description	Supply composition X_j^s (ppmw)	Target composition X_j^t (ppmw)	M_j	ε_j ppmw	C_j $/kg MSA	C_j^r $/kg CE removed
SW_1	Zeolite	3	15	0.09	15	0.70	58,333
SW_2	Air	0	10	0.10	100	0.05	5,000
SW_3	Steam	0	15	0.80	50	0.12	8,000

Component material balance for chloroethanol around the reactor
$(u = 1)$

Chloroethanol in recycled reactants

+ Chloroethanol generated due to oxychlorination reaction

= Chloroethanol in off-gas + Chloroethanol in wastewater W_6

+ Chloroethanol depleted by reduction reaction.   (7.12a)

Hence

$$W_5 z_5 + 6.03 \times 10^{-6} = V_1 y_1 + W_6 z_6 + 0.09 z_5. \qquad (7.12b)$$

But, as discussed in the problem statement, the compositions of CE in the gaseous and liquid effluents of the ethyl chloride reactor are related through an equilibrium distribution coefficient as follows:

$$\frac{y_1}{z_6} = 5. \qquad (7.13)$$

Therefore, Eq. (7.12b) can be further simplified by invoking Eq. (7.13) and substituting for the numerical values of V_1, W_5, and W_6 to get

$$0.180 y_1 - 0.060 z_5 = 6.03 \times 10^{-6} \qquad (7.14a)$$

with y_1 and z_5 in mass fraction units. Therefore, the previous equation can be rewritten as

$$0.180 y_1 - 0.060 z_5 = 6.030 \qquad (7.14b)$$

with y_1 and z_5 in units of ppmw.

**Component material balance for chloroethanol around first
Scrubber $(u = 2)$**

$$V_1 y_1 + W_1 z_1 = V_2 y_2 + W_2 z_2 \qquad (7.15a)$$

Substituting for the values of V_1, V_2, W_1, and W_2 and noting that z_1 is zero (fresh water), we get

$$2(y_1 - y_2) - z_2 = 0. \qquad (7.15b)$$

The Kremser equation can be used to model the scrubber:

$$NTP = \frac{\ln\left[\left(1 - \frac{HV_1}{W_1}\right)\left(\frac{y_1 - Hz_1}{y_2 - Hz_1}\right) + \frac{HV_1}{W_1}\right]}{\ln\left(\frac{W_1}{HV_1}\right)} \qquad (7.16a)$$

where Henry's coefficient $H = 0.1$ and NTP $= 1.3$. Therefore,

$$1.3 = \frac{\ln\left(0.8\dfrac{y_1}{y_2} + 0.2\right)}{\ln(5)}$$

i.e.,

$$y_2 = 0.10y_1. \tag{7.16b}$$

Similar to Eqs. (7.15b) and (7.16b), one can derive the following two equations for the second scrubber ($u = 3$):

$$2(y_2 - y_3) - z_4 = 0 \tag{7.17}$$

and

$$y_3 = 0.10y_2. \tag{7.18}$$

Component material balance for chloroethanol around the Mixer ($u = 4$)

$$W_2z_2 + W_4z_4 = W_5z_5 \tag{7.19a}$$

By plugging the values of W_1, W_2 and W_3 into Eq. (7.19a), we get:

$$z_2 + z_4 - 2z_5 = 0 \tag{7.19b}$$

Hence, the gaseous and liquid path diagram equations (7.13)–(7.19) can be summarized by the following model (with all compositions in ppmw):

For $u = 1$,

$$0.180y_1 - 6.030 = 0.060z_5$$

$$y_1 - 5z_6 = 0.$$

For $u = 2$,

$$2y_2 + z_2 = 2y_1 \tag{P7.1}$$

$$y_2 = 0.10y_1.$$

For $u = 3$,

$$2y_3 + z_4 = 2y_2$$

$$y_3 = 0.10y_2.$$

For $u = 4$,

$$2z_5 = z_2 + z_4.$$

As mentioned earlier, the gaseous path diagram for CE involves three unknown compositions (y_1, y_2, and y_3), whereas the liquid path diagram for CE has four unknown compositions (z_2, z_4, z_5, and z_6). The foregoing seven equations can be solved simultaneously to get these compositions. In terms of LINGO input, model (P7.1) can be written as follows:

```
model:
0.180*y1 - 0.060*z5 = 6.030;
y1 - 5*z6 = 0.0;
2*y2 + z2 - 2*y1 = 0.0;
y2 - 0.10*y1 = 0.0;
2*y3 + z4 - 2*y2 = 0.0;
y3 - 0.10*y2 = 0.0;
2*z5 - z2 - z4 = 0.0;
end
```

The solution to this model gives the following compositions (in ppmw CE) prior to interception:

$$y_1 = 50.0$$

$$y_2 = 5.0$$

$$y_3 = 0.5$$

$$z_2 = 90.0$$

$$z_4 = 9.0$$

$$z_5 = 49.5$$

$$z_6 = 10.0$$

Figure 7.8 is a schematic representation of the path diagram for the liquid sources.

As can be seen from Fig. 7.9, the objective of this problem is to move the terminal wastewater node ($w = 6$) to the targeted location designated by the X. The challenge, is how should the other nodes by moved to meet this target at minimum cost?

To incorporate interception, Fig. 7.10 is developed in a manner similar to that described by Fig. 7.6. Next, we use the criteria of thermodynamic feasibility and cost to prescreen interception strategies for each node. For instance, because air lies to the left of steam and has a lower cost (per kg CE removed), it is chosen in favor of steam. Nodes lying to the right of air ($w = 2, 5$) can be intercepted down to 10 ppmw CE using air stripping. Any interception of these nodes below 10 ppmw CE should be handled by air (up to 10 ppmw CE) followed by zeolite. Similarly, nodes $w = 4$ and 6 can be intercepted by zeolite down to 1.6 ppmw CE.

Similarly, the gaseous path diagram and MSAs are represented by Fig. 7.11. Because of thermodynamic infeasibility, oil cannot be used to intercept any node.

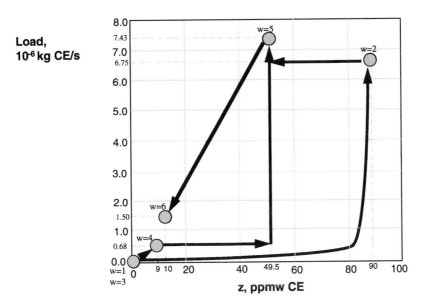

Figure 7.8 Liquid path diagram for CE case study.

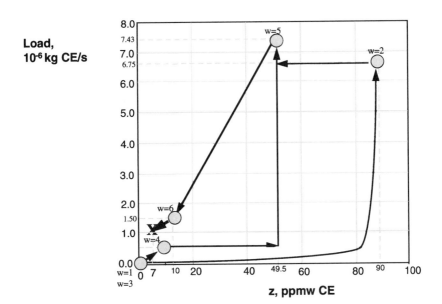

Figure 7.9 Liquid path diagram for CE case study with "X" marking waste-reduction target.

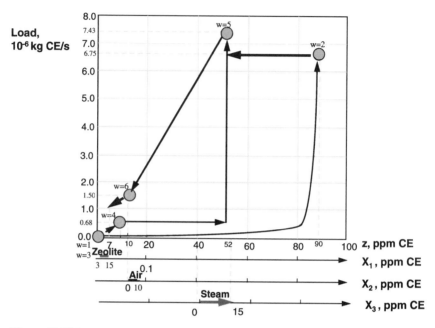

Figure 7.10 Pre-screening MSAs for intercepting liquid path diagram for CE case study.

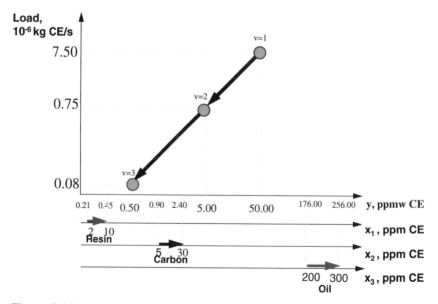

Figure 7.11 MSAs for intercepting gaseous path diagram for CE case study (schematic representation, not to scale).

Activated carbon should be used to intercept nodes $v = 1$ and 2 down to 0.9 ppmw CE. Any interceptions below 0.9 ppmw CE may be handled by the polymeric resin, which can reduce the CE content to 0.21 ppmw.

Having pre-screened the interception strategies, we should integrate these strategies with the path diagram as illustrated by Fig. 7.5. First, the path-diagram equations (P7.1) should be revised to include potential interception of all nodes as follows:

$$0.180y_1 - 6.030 = 0.060z_5^{int}$$

$$y_1 - 5z_6 = 0$$

$$2y_2 + z_2 = 2y_1^{int}$$

$$y_2 = 0.10y_1^{int} \qquad \text{(P7.2)}$$

$$2y_3 + z_4 = 2y_2^{int}$$

$$y_3 = 0.10y_2^{int}$$

$$2z_5 = z_2^{int} + z_4^{int}.$$

As part of the optimization procedure, we will have to determine which nodes should be intercepted and the optimal values of the intercepted nodes. Next, let us consider the case when only one[1] node is intercepted at a time.

Interception of $w = 2$: As determined earlier, air stripping is the optimal interception strategy for $w = 2$. Therefore, the following LINGO model can be written to minimize the stripping cost[2]:

```
model:
min = 0.05*3600*8760*Lair;
0.180*y1 - 0.060*z5 = 6.030;
y1 - 5*z6 = 0.0;
2*y2 + z2 - 2*y1 = 0.0;
y2 - 0.10*y1 = 0.0;
2*y3 + z4 - 2*y2 = 0.0;
y3 - 0.10*y2 = 0.0;
2*z5 - zINT2 - z4 = 0.0;
```

---

[1] Simultaneous interception of multiple sources may be considered by allowing intercepted compositions of multiple nodes to be treated as optimization variables in the same formulation. See El-Halwagi *et al.* (1996) for more details.

[2] Here, the objective is to minimize operating cost followed by trading off with fixed cost as discussed in Chapter Two. If needed, one can directly minimize total annualized cost by adding a fixed-cost term to the objective function. This fixed-cost term is expressed in terms of z_2^{int} as described in Chapter Two.

```
z6 <= 7.0;
10*Lair-0.075*(z2 - zINT2) = 0.0;
end
```

The following is a LINGO output summarizing the solution:

```
Objective value:          713108
          Variable        Value
          LAIR            0.45225
          Y1              35.00000
          Z5              4.499997
          Z6              7.000000
          Y2              3.500000
          Z2              63.00000
          Y3              0.3500000
          Z4              6.300000
          ZINT2           2.699993
```

The solution indicates that to reduce the CE content of the terminal wastewater stream to 7 ppmw, air stripping can be used to intercept the bottom product of the first scrubber ($w = 2$) to 2.7 ppmw CE at a cost of 713,108/yr. Since air stripping can only reduce composition to 10 ppm, zeolite should be used to remove CE from 10 ppm to 2.7 ppm. Therefore, the cost should be adjusted to \$1,633,960/yr. More details on this adjustment will be given when $w = 5$ is intercepted. The same procedure can be repeated for the other nodes as shown below.

Interception of $w = 4$:

model:
```
min = 0.70*3600*8760*LZeolite;
0.180*y1 - 0.060*z5 = 6.030;
y1 - 5*z6 = 0.0;
2*y2 + z2 - 2*y1 = 0.0;
y2 - 0.10*y1 = 0.0;
2*y3 + z4 - 2*y2 = 0.0;
y3 - 0.10*y2 = 0.0;
2*z5 - z2 - zINT4 = 0.0;
z6 <= 7.0;
12*LZeolite-0.075*(z4 - zINT4)=0.0;
end
```

```
Solution: INFEASIBLE
```

Since no feasible solution was found, there is no way that $w = 4$ can be intercepted to reduce the CE content of the terminal wastewater stream to 7 ppmw. Indeed, even if CE in $w = 4$ is completely eliminated, the terminal wastewater stream will drop its content of CE to only 9.6 ppmw, as shown by the following model:

model:
```
0.180*y1 - 0.060*z5 = 6.030;
y1 - 5*z6 = 0.0;
2*y2 + z2 - 2*y1 = 0.0;
y2 - 0.10*y1 = 0.0;
2*y3 + z4 - 2*y2 = 0.0;
y3 - 0.10*y2 = 0.0;
2*z5 - z2 - zINT4 = 0.0;
zINT4 = 0.0;
end
```

Variable	Value
Y1	47.85714
Z5	43.07143
Z6	9.571429
Y2	4.785714
Z2	86.14286
Y3	0.4785714
Z4	8.614286
ZINT4	0.0000000E+00

Interception of $w = 5$:

model:
```
min = 0.05*3600*8760*Lair;
0.180*y1 - 0.060*zINT5 = 6.030;
y1 - 5*z6 = 0.0;
2*y2 + z2 - 2*y1 = 0.0;
y2 - 0.10*y1 = 0.0;
2*y3 + z4 - 2*y2 = 0.0;
y3 - 0.10*y2 = 0.0;
2*z5 - z2 - z4 = 0.0;
Z6 <= 7.0;
10*Lair-0.15*(z5 - zINT5)=0.0;
end
```

Objective value: 703108

Variable	Value
LAIR	0.45225.
Y1	35.00000

```
ZINT5          4.499997
  Z6           7.000000
  Y2           3.500000
  Z2           63.00000
  Y3           0.3500000
  Z4           6.300000
  Z5           34.65000
```

Since z_5 was intercepted to 4.5 ppmw, the separation task cannot be done by air alone (air can reduce the composition to 10 ppmw). Hence, the program is adjusted to include zeolite.

model:
```
min = 0.05*3600*8760*Lair + 0.70*3600*7860*LZeolite;
0.180*y1 - 0.060*zINT5 = 6.030;
y1 - 5*z6 = 0.0;
2*y2 + z2 - 2*y1 = 0.0;
y2 - 0.10*y1 = 0.0;
2*y3 + z4 - 2*y2 = 0.0;
y3 - 0.10*y2 = 0.0;
2*z5 - z2 - z4 = 0.0;
Z6 <= 7.0;
10*Lair-0.15*(z5 - 10.0) = 0.0;
12*LZeolite-0.15*(10.0 - zINT5) = 0.0;
end
```

```
Objective value:          2100693.
              Variable     Value
              LAIR         0.5925000
              LZEOLITE     0.6875004E-01
              Y1           35.00000
              ZINT5        4.499997
              Z6           7.000000
              Y2           3.500000
              Z2           63.00000
              Y3           0.3500000
              Z4           6.300000
              Z5           34.65000
```

Interception of $w = 6$ (Terminal Wastewater):

model:
```
min = 0.70*3600*8760*LZeolite;
0.180*y1 - 0.060*z5 = 6.030;
```

```
y1 - 5*z6 = 0.0;
2*y2 + z2 - 2*y1 = 0.0;
y2 - 0.10*y1 = 0.0;
2*y3 + z4 - 2*y2 = 0.0;
y3 - 0.10*y2 = 0.0;
2*z5 - z2 - z4 = 0.0;
zINT6 <= 7.0;
12*LZeolite-0.150*(10.0 - zINT6) = 0.0;
end
```

```
Objective value:           827820.0

                Variable       Value
                LZEOLITE       0.3750000E-01
                      Y1       50.00000
                      Z5       49.50000
                      Z6       10.00000
                      Y2       5.000000
                      Z2       90.00000
                      Y3       0.5000000
                      Z4       9.000000
                   ZINT6       7.000000
```

Alternatively,

model:
```
min = 0.70*3600*8760*LZeolite;
zINT6 <= 7.0;
12*LZeolite-0.150*(10.0 - zINT6)=0.0;
end
```

```
Optimal solution found at step:   0
Objective value:           827820.0

                Variable       Value
                LZEOLITE       0.3750000E-01
                   ZINT6       7.000000
```

Interception of $v = 1$:

model:
```
min = 0.10*3600*8760*LCarbon;
0.180*y1 - 0.060*z5 = 6.030;
```

```
y1 - 5*z6 = 0.0;
2*y2 + z2 - 2*yINT1 = 0.0;
y2 - 0.10*yINT1 = 0.0;
2*y3 + z4 - 2*y2 = 0.0;
y3 - 0.10*y2 = 0.0;
2*z5 - z2 - z4 = 0.0;
z6 <= 7.0;
25*LCarbon-0.150*(y1 - yINT1) = 0.0;
end
```

```
Optimal solution found at step:    1
Objective value:            576248.8
```

Variable	Value
LCARBON	0.1827273
Y1	35.00000
Z5	4.499997
Z6	7.000000
Y2	0.4545451
Z2	78.181812
YINT1	4.545451
Y3	0.4545451E-01
Z4	0.8181812

Interception of $v = 2$:

model:
```
min = 0.10*3600*8760*LCarbon;
0.180*y1 - 0.060*z5 = 6.030;
y1 - 5*z6 = 0.0;
2*y2 + z2 - 2*y1 = 0.0;
y2 - 0.10*y1 = 0.0;
2*y3 + z4 - 2*yINT2 = 0.0;
y3 - 0.10*yINT2 = 0.0;
2*z5 - z2 - z4 = 0.0;
z6 <= 7.0;
25*LCarbon-0.150*(y2 - yINT2) = 0.0;
end
```

```
INFEASIBLE
```

[This is a situation similar to the interception of $w = 4$].

Based on these results, the optimal single interception for the problem is to use activated-carbon adsorption to separate CE from the gaseous stream leaving the reactor ($v = 1$) and reduce its composition to $y_1^{int} = 4.55$ ppmw CE (which corresponds to removing 4.57×10^{-6} kg CE/s from $v = 1$). The optimal solution has a minimum operating cost of approximately $ 576,250/yr. Several important observations can be drawn from the list of generated solutions:

- Waste reduction is a multimedia (multiphase) problem. The solution to a wastewater problem may lie in the gas phase.
- In-plant interception may be superior to terminal-waste separation. For instance, separating CE from the terminal wastewater stream incurs an annual operating cost of $827,820/year, which is 44% more expensive than the optimal solution.
- More mass may be removed for less cost. For instance, intercepting $v = 1$ involves the removal of 4.57×10^{-6} kg CE/s (compared to the before-interception data) while the more expensive interception of $w = 6$ involves the removal of 0.45×10^{-6} kg CE/s. This result can be explained based on thermodynamic and mass-transfer arguments. Removal of the pollutant from more concentrated streams is more thermodynamically favorable than removal from less concentrated streams. Therefore, it is typically less expensive to use low-cost MSAs to eliminate the pollutant from in-plant streams than it is to employ high-cost MSAs to remove traces of pollutant from the dilute terminal streams. It is also worth noting the nonlinear propagation of mass of the pollutant throughout the process. For a given effect on the pollutant content of a terminal stream, different masses may be removed from different in-plant streams. Another reason for this mass-cost observation is that removal of the pollutant in one medium can be less expensive than separation from another medium. For instance, a certain amount of the pollutant may be removed from air at a lower cost than that of recovering a smaller quantity from wastewater.

7.4 Developing Strategies for Segregation, Mixing, and Direct Recycle

The previous sections have focused on in-plant interception for waste reduction. In this section, we target other mass-integration strategies, including stream segregation, mixing, and direct recycle (without interception). Later in this chapter, interception strategies will be incorporated. Without interception, Fig. 7.1 is revised to the structural representation shown in Fig. 7.12. This representation embeds potential configurations of interest by allowing each source to be segregated, mixed, allocated to a unit, and recycled back to the process. The optimization task is to determine the flowrate and composition of each stream in Fig. 7.12 while

SOURCES   SINKS/   SOURCES
                  GENERATORS   (Back to Process)

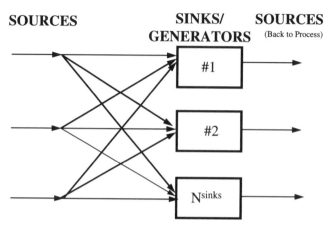

Figure 7.12 Structural representation of segregation, mixing and direct recycle options.

considering the path-diagram expressions to incorporate the interactions among units and streams. This task can be mathematically formulated as shown in the following section.

7.5 Case Study Revisited: Segregation, Mixing, and Recycle for the Chloroethanol Case Study

The scope of the previously addressed CE case study is now altered to allow for stream segregation, mixing, and recycle within the ethyl chloride plant. There are five sinks: the reactor ($u = 1$), the first scrubber ($u = 2$), the second scrubber ($u = 3$), the mixing tank ($u = 4$) and the biotreatment facility for effluent treatment ($u = 5$). There are six sources of CE-laden aqueous streams ($w = 1\text{--}6$). There is the potential for segregating two liquid sources ($w = 2, 4$). The following process constraints should be considered:

Composition of aqueous feed to reactor	≤ 65 ppmw CE	(7.20)
Composition of aqueous feed to first scrubber	≤ 8 ppmw CE	(7.21)
Composition of aqueous feed to second scrubber	$= 0$ ppmw CE	(7.22)
$0.090 \leq$ Flowrate of aqueous feed to reactor (kg/s)	≤ 0.150	(7.23)
$0.075 \leq$ Flowrate of aqueous feed to first scrubber (kg/s)	≤ 0.090	(7.24)
$0.075 \leq$ Flowrate of aqueous feed to second scrubber (kg/s)	$\leq 0.085.$	(7.25)

The objective of this case study is to determine the target for minimizing the total load (flowrate $\times$ composition) of CE discharged in terminal wastewater of the plant using segregation, mixing, and recycle strategies.

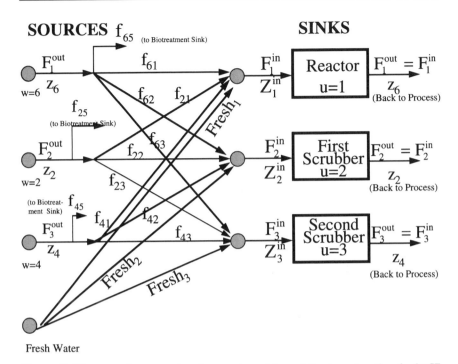

Figure 7.13 Structural representation of segregation, mixing and direct recycle options for the CE case study.

Solution

Figure 7.13 is structural representation of segregation, mixing, and direct recycle candidate strategies for the problem. Each source is split into several fractions that can be fed to a sink. The flowrate of the streams passed from source w to sink u is referred to as f_{wu}. The terms F_u^{in}, Z_u^{in}, and F_u^{out} represent the inlet flowrate, inlet composition, and outlet flowrate of the streams associated with unit u. Since mixing is embedded, there is no need to include the mixing tank ($u = 4$) or the source that it generates ($w = 5$) in the analysis. Unless recycle of biotreatment effluent is considered, there is no need to represent the biotreatment sink in Fig. 7.13. However, streams allocated to biotreatment should be represented and their flowrates are referred to as f_{w5} ($u = 5$ is the biotreatment sink). Finally, fresh water may be used in any unit u at a flowrate of $Fresh_u$.

Since the flowrates of the streams fed to the sinks are to be determined as part of the optimization problem, the path equations represented by model (P7.1) should be revised as follows to allow for variable flowrate of the streams:

$$\left(0.15 + 0.2^{*}F_1^{in}\right)y_1 - \left(F_1^{in} - 0.09\right)Z_1^{in} = 6.030$$

$$0.15(y_1 - y_2) - F_2^{in}\left(z_2 - Z_2^{in}\right) = 0.0$$

$$\left(\frac{F_2^{in}}{0.015}\right)^{1.3} - \frac{\left(1 - \dfrac{0.015}{F_2^{in}}\right)\left(y_1 - 0.1Z_2^{in}\right)}{y_2 - 0.1Z_2^{in}} - \frac{0.015}{F_2^{in}} = 0.00 \qquad \text{(P7.3)}$$

$$0.15(y_2 - y_3) - F_3^{in}\left(z_4 - Z_3^{in}\right) = 0.0$$

$$\left(\frac{F_3^{in}}{0.015}\right)^{1.3} - \frac{\left(1 - \dfrac{0.015}{F_3^{in}}\right)\left(y_2 - 0.1Z_3^{in}\right)}{y_3 - 0.1Z_3^{in}} - \frac{0.015}{F_3^{in}} = 0.0.$$

The revised path diagram is integrated with material allocation equations to form the constraints for the mathematical formulation. The following model presents the optimization program as a LINGO file. The commented-out lines (preceded by !) are explanatory statements that are not part of the formulation.

model:

```
! Objective is to minimize load of CE in terminal
  wastewater streams
min = F65*z6 + F25*z2 + F45*z4;

! Water balances around inlets of sinks
F1In - f21 - f41 - f61 - Fresh1 = 0.0000;
F2In - f22 - f42 - f62 - Fresh2 = 0.0000;
F3In - f23 - f43 - f63 - Fresh3 = 0.0000;

! CE balances around inlets of sinks
F1In*Z1In - f21*z2 - f41*z4 - f61*z6 = 0.0000;
F2In*Z2In - f22*z2 - f42*z4 - f62*z6 = 0.0000;
F3In*Z3In - f23*z2 - f43*z4 - f63*z6 = 0.0000;

! Water balances around sinks
F1In - F1Out = 0.0000;
F2In - F2Out = 0.0000;
F3In - F3Out = 0.0000;

! Water balances for sources to be split
F1Out - f65 - f61 - f62 - f63 = 0.0000;
```

```
F2Out - f25 - f21 - f22 - f23 = 0.0000;
F3Out - f45 - f41 - f42 - f43 = 0.0000;

! Path-diagram equations
(0.15 + 0.2*F1In)*y1 - (F1In - 0.09)*Z1In = 6.030;
0.15*(y1 - y2) - F2In*(z2 - Z2In) = 0.0000;
(F2In/0.015)^ 1.3 - (1 - 0.015/F2In)*(y1 - 0.1*Z2In)/
   (y2 - 0.1*Z2In) - 0.015/F2In = 0.0000;
0.15*(y2 - y3) - F3In*(z4 - Z3In) = 0.0000;
(F3In/0.015)^ 1.3 - (1 - 0.015/F3In)*(y2 - 0.1*Z3In)/
   (y3 - 0.1*Z3In) - 0.015/F3In = 0.0;
   y1 - 5*z6 = 0.0000;

! Restrictions on what can be recycled
Z1In <= 65.0000;
Z2In < 8.0000;
Z3In = 0.0000;
F1In <= 0.1500;
F1In >= 0.0900;
F2In <= 0.0900;
F2In >= 0.07500;
F3In <= 0.0850;
F3In >= 0.0750;

! If solution gives negative flows or compositions,
   the non-negativity constraints can then
! be added
end
```

This is a nonlinear program which can be solved using LINGO to get the following answer:

```
Objective value:        0.4881247

            Variable       Value
              F65          0.0000000E+00
              Z6           7.178572
              F25          0.0000000E+00
              Z2           60.96842
              F45          0.7500000E-01
              Z4           6.508329
              F1IN         0.9000000E-01
```

F21	0.9000000E-01
F41	0.0000000E+00
F61	0.0000000E+00
FRESH1	0.0000000E+00
F2IN	0.9000000E-01
F22	0.0000000E+00
F42	0.0000000E+00
F62	0.9000000E-01
FRESH2	0.0000000E+00
F3IN	0.7500000E-01
F23	0.0000000E+00
F43	0.0000000E+00
F63	0.0000000E+00
FRESH3	0.7500000E-01
Z1IN	60.96842
Z2IN	7.178572
Z3IN	0.0000000E+00
F1OUT	0.9000000E-01
F2OUT	0.9000000E-01
F3OUT	0.7500000E-01
Y1	35.89286
Y2	3.618950
Y3	0.3663237

These results can be used to construct the solution as shown in Fig. 7.14. The target for minimum CE discharge through segregation, mixing and direct recycle is 0.488×10^{-6} kg/s (about 15 kg/yr). The solution indicates that the optimal policy is to segregate the effluents of the two scrubbers, pass the effluent of the first scrubber to the reactor, recycle the aqueous effluent of the reactor to the first scrubber and dispose of the second scrubber effluent as the terminal wastewater stream.

Note that this case study has presented mass integration opportunities via intraprocess (within the ethyl chloride plant) integration. The scope of the case study can be broadened to incorporate interprocess (e.g., between the ethanol plant and the ethyl chloride plant) integration. For instance, the wastewater leaving the distillation column (0.030 kg/s) does not contain any CE. If the composition of the various species contained in this wastewater is also within tolerated recycle limits for the second scrubber, it can be used to replace some of the fresh water. This solution is shown in Fig. 7.15. Once these integration opportunities have been identified, a cost-benefit analysis should be undertaken to determine the applicability of these strategies.

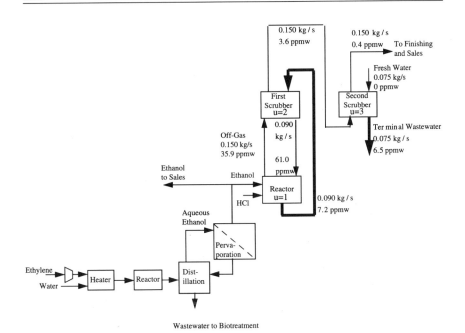

Figure 7.14 Solution to the CE case study using segregation, mixing and direct recycle.

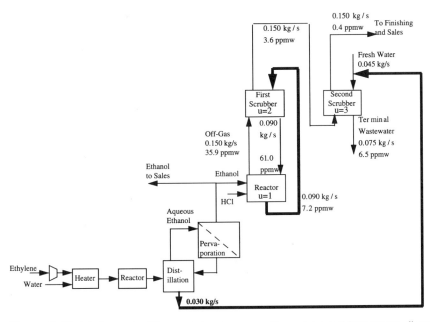

Figure 7.15 Solution to the CE case study using segregation, mixing and inter-process direct recycle.

7.6 Integration of Interception with Segregation, Mixing, and Recycle

The procedures presented heretofore can be combined to allow the integration of interception with other mass-integration strategies. For instance, suppose that we wish to use interception to cut the discharged load of CE in Fig. 7.15 by 50% by seeking a target composition of 3.3 ppmw CE in the aqueous effluent of the second scrubber (discharged load $= 0.25 \times 10^{-6}$ kg CE/s). Which nodes should be intercepted with which separating agents to meet the target at minimum cost? The procedure presented in Sections 7.2 and 7.3 can be directly applied to yield an optimal solution (Fig. 7.15) that uses activated carbon to intercept the gaseous stream leaving the first scrubber ($v = 2$) and reduce its composition to $y_2^{int} =$ 1.83 *ppmw*. The annual operating cost of the system is approximately \$32,000/yr. This interception propagates through the process, resulting in a terminal wastewater composition of 3.3 ppmw CE.

It is interesting to compare the optimal configuration shown in Fig. 7.16 to end-of-pipe solutions. Suppose we retain the identified segregation, mixing, and direct recycle strategies shown in Fig. 7.14, but intercept the wrong stream. For instance, if the CE content of the terminal wastewater stream is to be reduced from 6.5 ppmw

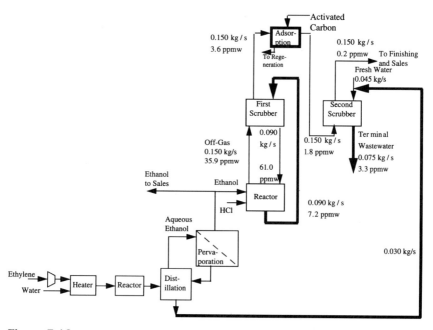

Figure 7.16 Solution to the CE case study using segregation, mixing, interception and recycle.

to 3.3 ppmw, the cheapest practically feasible MSA is Zeolite (this choice can be made based on Fig. 7.10). This end-of-pipe separation (after segregation, mixing, and recycle) requires an annual operating cost of \$441,500/yr which is 14 times more expensive than the optimal solution. If end-of-pipe separation is employed without segregation, mixing, and recycle, the targeted disposal limit of 0.25×10^{-6} kg CE/s can be met by reducing the CE content of the reactor effluent (the original wastewater stream of Fig. 7.7) from 10.0 to 1.8 ppmw CE. Again, zeolite is used for this separation based on Fig. 7.10. The operating cost associated with this end-of-pipe solution is about \$2.3 million/yr. These nonoptimal options meet the CE-loading target and provide the same waste-reduction effect as the optimal solution. However, as can be seen from this case study, the costs of these solutions can be significantly higher than those of solutions obtained using mass integration.

Problems

7.1 Solve problem 4.1. using mathematical optimization.

7.2 Employing mathematical programming, solve problem 4.2.

7.3 Consider the kraft pulp and paper process (Parthasarathy *et al.*, 1996; Lovelady *et al.*, 1996) shown by Fig. 7.17. A description of the pulping portion process was given in problem 4.3. Several aqueous effluents are disposed off the plant including wastewater from the bleaching plant, brown stock washers, and the evaporator. The relevant material balance data are given in Fig. 7.17. In order to partially close the water loop of the process and reduce wastewater, several aqueous streams must be recycled. The difficulty involved in water recycling is that certain ionic species such as chlorine, potassium, silicon and sodium tend to accumulate in the process as partial closure is carried out. This leads to problems such as corrosion and plugging.

The objective of this problem is to employ mass integration techniques to reduce wastewater discharge while alleviating any buildup of ionic species. In this problem, we focus our attention on chlorine ions as the key species.

The following process constraints are given for the sinks :

Washers/Filter :
$2.0 \leq$ Wash Feed flowrate (kg/s) ≤ 3.0
$0.0 \leq$ chlorine concentration (ppmw) ≤ 2

Bleaching Plant :
$5.0 \leq$ Water Flowrate (kg/s) ≤ 5.5
$20 \leq$ chlorine concentration (ppmw) ≤ 30

Two external MSAs may be considered for removal of chlorine ions: activated carbon (S_1) and ion exchange resin (S_2). The equilibrium relation for removing chlorine by these MSAs is given by:
$$y = m_j x_j$$
where y and x_j are in units of ppmw. Data for the MSAs are given in Table 7.3.

Table 7.3
Data for MSAs That Can Remove Chlorine from Liquid Streams

Stream	Supply composition (ppmw)	Target composition (ppmw)	m_j	ε_j ppmw	cost $/kg MSA
S_1	5	10	2.0	0.1	0.0010
S_2	6	16	0.5	2.0	0.0015

7.4 Consider the tricresyl phosphate process shown in Fig. 7.18. Tricresyl phosphate, $(CH_3C_6H_4O)_3PO$, can be produced by reacting cresol with phosphorus oxychloride. First, phosphorus oxychloride and creosol are mixed and heated then fed continuously and slowly to a reactor. Inert gas is added to the reactor as a diluent. The exit product stream from the reactor is cooled and washed twice to improve the purity of the product. Washing the product results in reducing the concentration of cresol from 500 ppm to .007 ppm in the final product. To insure proper wetting, the mass flowrate of washing water in the first and second washing units (W_1 and W_2) should be at least seven times the mass flowrate of the product stream. The water streams used in the first and second washing units result in wastewater streams W_3 and W_4 which are mixed to result in a terminal wastewater stream (W_5).

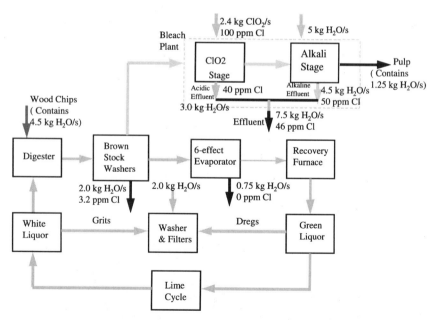

Figure 7.17 Pulping and bleaching process (Parthasarathy *et al.*, 1996).

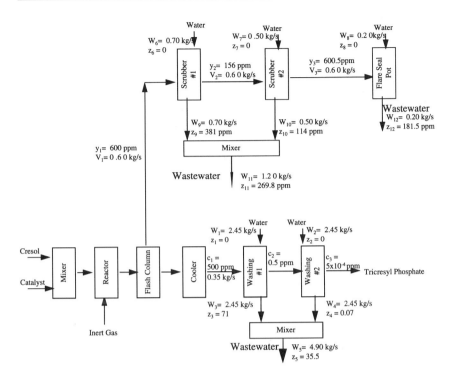

Figure 7.18 Tricresyl phosphate process (Hamad *et al.*, 1996).

The following constraints are imposed on the quality and quantity of washing water:
Constraints on flowrate of washing water:

First washing unit: $W_1 = 2.45$ kg/s
Second washing unit: $W_2 = 2.45$ kg/s

Constraints on concentration of cresol in washing water:
First washing unit: $0 \leq z_1 \leq 5$ ppm
Second washing unit: $z_2 = 0$ ppm (fresh water must be used for washing)

The gaseous stream exiting the reactor has a flowrate of 0.6 kg/s and contains 600 ppm cresol. It passes through two water scrubbers before it goes finally to a flare system for burning. To prevent backward propagation of fire in the flare, water is utilized in a water valve to seal the flare. Water utilized in scrubber #1 (W_6), scrubber #2 (W_7), and flare seal pot (W_8) possess the following constraints:

Constraints on feed water flowrate:

First scrubber: $0.7 \leq W_6 \leq 0.84$, kg/s
Second scrubber: $0.5 \leq W_7 \leq 0.6$, kg/s

Flare seal pot: $0.2 \leq W_8 \leq 0.25$, kg/s

Constraints on the concentration of cresol in feed water:
First scrubber: $0. \leq z_6 \leq 30$ ppm
Second scrubber: $0 \leq z_7 \leq 30$ ppm
Flare seal pot: $0 \leq z_8 \leq 100$ ppm

Wastewater streams W_9 and W_{10} are mixed to form a terminal wastewater stream W_{11}. Terminal wastewater stream W_{12} is generated in the flare system.

The concentration of cresol in the product stream is denoted by c in Fig. 7.18. The following modeling correlations describe the performance of the involved units:

For first washing unit (u = 1):

$$c_2 = .001^*c_1 + .8^*z_1$$

$$0.34^*(c_1 - c_2) - W_1^*(z_3 - z_1) = 0$$

For second washing unit (u = 2):

$$c_3 = .001^*c_2 + .8^*z_2$$

$$0.34^*(c_2 - c_3) - W_1^*(z_4 - z_2) = 0$$

For first scrubber (u = 3):

$$0.45 = \frac{\ln\left[\left(1 - \dfrac{HV_1}{W_6}\right)\left(\dfrac{y_1^{int} - Hz_6}{y_2 - Hz_6}\right) + \dfrac{HV_1}{W_6}\right]}{\ln\left(\dfrac{W_6}{HV_1}\right)}$$

For second scrubber (u = 4):

$$0.35 = \frac{\ln\left[\left(1 - \dfrac{HV_2}{W_7}\right)\left(\dfrac{y_2^{int} - Hz_7}{y_3 - Hz_7}\right) + \dfrac{HV_2}{W_7}\right]}{\ln\left(\dfrac{W_7}{HV_2}\right)}$$

For flare seal pot (u = 5):

$$5.4 = \frac{\ln\left[\left(1 - \dfrac{HV_3}{W_8}\right)\left(\dfrac{y_3^{int} - Hz_8}{y_4 - Hz_8}\right) + \dfrac{HV_3}{W_8}\right]}{\ln\left(\dfrac{W_8}{HV_3}\right)}$$

where y_4 is the concentration of the gaseous stream after passing through the water seal pot in the flare system, and Henry's coefficient (H) is 0.064 (in units of weight fraction by wieght fraction).

Cresol is a toxic chemical that can be absorbed via the skin and may cause damage to the kidney, the liver, and nervous system. The objective of this problem is to reduce cresol concentration in any discharged wastewater stream to 5 ppmw or less.

Table 7.4
Data for MSAs That Can Remove Cresol from Liquid Streams

Stream	Description	Supply composition (ppmw)	Target composition (ppmw)	m_j	ε_j ppmw	Cost $/kg MSA
SW_1	Oil	0	343	0.0584	685	0.04
SW_2	Light gas	15	61	0.108	13	0.02

The following data may be used to solve the problem.

Interception Techniques:

To reduce cresol content in discharged wastewater streams, four MSAs are available: two MSAs to remove cresol from gaseous streams and two MSAs to remove cresol from liquid streams. Tables 7.4. and 7.5. provide data for the candidate MSAs.

Piping and Pumping Cost:

The cost for piping and pumping is a assumed to be linear with the mass flowrate of the stream. The cost of piping (c_{pipe}) is approximated by:

$$c_{pipe} = \$3.3 \times 10^{-4}/(m \cdot kg/s)$$

The distances among units needed for piping are provided in Table 7.6.

Equipment Cost Data:

The mass-exchange columns are assumed to be packed and have a depreciation period of three years. One-meter diameter columns are used. The following cost may be used:

$$\text{cost of packing (\$)} = 5,250/m^3 \cdot year$$

$$\text{cost of column (\$)} = 12,600^* \text{ height of column shell,}$$

where

$$\text{height of column shell} = 1.3^* \text{ tray spacing}^* \text{number of trays.}$$

Table 7.5
Data for MSAs That Can Remove Cresol from Gaseous Streams

Stream	Description	Supply composition (ppmw)	Target composition (ppmw)	m_j	ε_j ppmw	cost $/kg MSA
SV_1	Polymeric resin	0	1023	0.185	1135	0.03
SV_2	Adsorbent	4	560	0.027	1292	0.02

Table 7.6
Distances among Units

Source	Sink	Distance, m
Washing Unit # 2 (W_4)	Washing Unit # 1	30
Washing Unit # 2 (W_4)	Scrubber # 1	100
Washing Unit # 2 (W_4)	Scrubber # 2	110
Washing Unit # 2 (W_4)	Flare Seal Pot	90
Washing Unit # 1 (W_3)	Scrubber # 1	80
Washing Unit # 1 (W_3)	Scrubber # 2	90
Washing Unit # 1 (W_3)	Flare Seal Pot	70
Washing Unit # 1 (W_3)	Mass-Exchanger*	20
Scrubber # 1 (W_9)	Scrubber # 2	20
Scrubber # 1 (W_9)	Flare Seal Pot	95
Scrubber # 1 (W_9)	Mass-Exchanger*	20
Scrubber # 2 (W_{10})	Flare Seal Pot	75
Scrubber # 2 (W_{10})	Mass-Exchanger*	20
Flare Seal Pot (W_{12})	Mass-Exchanger*	80

*In case additional mass exchangers are added to the plant to tackle the streams leaving these units.

A tray spacing of 0.41 m may be employed. All columns are assumed to have an overall efficiency of 25%. The fixed cost of the column is obtained by multiplying the equipment cost by a factor of 5 to account for installation, instrumentation and other ancillary devices.

Symbols

b_j	intercept of equilibrium line for mass separating agent j
f_{wu}	flowrate of stream allocated from source w to unit u, kg/s
F_u^{in}	input flowrate to unit u, kg/s
F_u^{out}	output flowrate from unit u, kg/s
Fresh$_u$	fresh water flowrate to unit u, kg/s
H	Henry's coefficient
l_j	mass flowrate of mass separating agent j for pollutant-laden gaseous streams, kg/s
L_j	mass flowrate of mass separating agent j for pollutant-laden liquid streams, kg/s
m_j	slope of equilibrium line for mass separating agent j for the gas phase
M_j	slope of equilibrium line for mass separating agent j for the liquid phase
NTP	number of theoretical stages
u	index for pollutant-processing unit (sink/generator)
v	index for pollutant-laden gaseous stream (source)
V_v	flowrate of gaseous stream (source) v, kg/s
w	index for pollutant-laden liquid stream (source)

W_w	flowrate of liquid source w, kg/s
x_j	mass fraction of pollutant in mass separating agent j
x_j^s	supply composition of pollutant in mass separating agent j for pollutant-laden gaseous streams
x_j^t	target composition of mass separating agent j for pollutant-laden gaseous streams
X_j^s	supply composition of pollutant in mass separating agent j for pollutant-laden liquid streams
X_j^t	target composition of pollutant in mass separating agent j for pollutant-laden liquid streams
y_v	composition of pollutant in gaseous stream v
y_v^{int}	composition of pollutant in gaseous stream v after interception
$z_i^{Terminal}$	composition of pollutant in terminal liquid waste stream i
Z_u^{in}	pollutant composition entering unit u
z_w	composition of pollutant in liquid stream w
z_w^{int}	composition of pollutant in liquid stream w after interception

Subscripts

i	waste stream i
j	mass separating agent j
u	pollutant-processing (sink/generator) unit u
v	pollutant-laden gaseous stream v
w	pollutant-laden liquid stream w

Superscripts

in	inlet
int	intercepted
out	outlet
s	supply
t	target
Terminal	denotes terminal streams

Greek Letters

ε_j	minimum allowable composition for mass separating agent j between equilibrium and operating lines
ϕ_a	mass generated of pollutant in the reactor
ϕ_v	mass load of pollutant in gaseous stream v
ψ_w	mass load of pollutant in liquid stream w

References

El-Halwagi, M. M., Hamad, A. A., and Garrison, G. W. (1996). Synthesis of waste interception and allocation networks. *AIChE J.*, **42**(11), 3087–3101.

El-Halwagi, M. M. and Spriggs, H. D. (1996). An integrated approach to cost and energy efficient pollution prevention. Proceedings of Fifth World Congr. of Chem. Eng., Vol. III, pp. 344–349, San Diego.

Garrison, G. W., Hamad, A. A., and El-Halwagi, M. M. (1995). Synthesis of waste interception networks. *AIChE Annu. Meet.*, Miami.

Garrison, G. W., Spriggs, H. D., and El-Halwagi, M. M. (1996). A global approach to integrating environmental, energy, economic and technological objectives. Proceedings of Fifth World Congr. of Chem. Eng., Vol. I, pp. 675–680, San Diego.

Hamad, A. A., Garrison, G. W., Crabtree, E. W. and El-Halwagi, M. M. (1996). Optimal design of hybrid separation systems for waste reduction. Proceedings of Fifth World Congr. of Chem. Eng., Vol. III, pp. 453–458, San Diego.

Hamad, A. A., Varma, V., El-Halwagi, M. M., and Krishnagopalan, G. (1995). Systematic integration of source reduction and recycle reuse for the cost-effective compliance with the cluster rules. *AIChE Annu. Meet.*, Miami.

Lovelady, E., Parthasarathy, G., Krishnagopalan, G., and El-Halwagi, M.M. (1996). Application of Mass Integration Techniques Towards Systematic and Economical Waste Minimization. Proceedings of the ACS Ind. Eng. Chem. Special Symposium, Emerging Technologies in Hazardous Waste Management VIII, pp. 379-382, ACS, Birmingham, AL.

Parthasarathy, G., Lovelady, E., Hamad, A.A., Krishnagopalan, G., and El-Halwagi, M.M. (1996). Application of Mass Integration for the Cost-Effective Pollution Prevention in Pulp and Paper Mills. *AIChE Annu. Meet.*, Chicago.

CHAPTER EIGHT

Synthesis of Reactive
Mass-Exchange Networks

Chapters Three, Five and Six have covered the synthesis of physical mass-exchange networks. In these systems, the targeted species were transferred from the rich phase to the lean phase in an intact molecular form. In some cases, it may be advantageous to convert the transferred species into other compounds using reactive MSAs. Typically, reactive MSAs have a greater capacity and selectivity to remove an undesirable component than physical MSAs. Furthermore, since they react with the undesirable species, it may be possible to convert pollutants into other species that may either be reused within the plant itself or sold.

The synthesis of a network of reactive mass exchangers involves the same challenges described in synthesizing physical MENs. The problem is further compounded by virtue of the reactivity of the MSAs. Driven by the need to address this important problem, Srinivas and El-Halwagi introduced the problem of synthesizing *reactive mass-exchange networks* REAMENs and developed systematic techniques for its solution (Srinivas and El-Halwagi, 1994; El-Halwagi and Srinivas, 1992). This chapter provides the basic principles of synthesizing REAMENs. The necessary thermodynamic concepts are covered. Chemical equilibrium is then tackled in a manner that renders the REAMEN synthesis task close to the MEN problem. Finally, an optimization-based approach is presented and illustrated by a case study.

8.1 Objectives of REAMEN Synthesis

The problem of synthesizing REAMENs can be stated as follows (El-Halwagi and Srinivas, 1992):

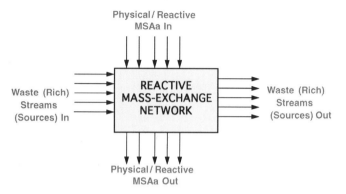

Figure 8.1 Schematic representation of the REAMEN synthesis problem.

Given a number N_R of waste (rich) streams and a number N_S of lean streams (physical and reactive MSAs), it is desired to synthesize a cost-effective network of physical and/or reactive mass exchangers which can preferentially transfer a certain undesirable species, A, from the waste streams to the MSAs whereby it may be reacted into other species. Given also are the flowrate of each waste stream, G_i, its supply (inlet) composition, y_i^s, and target (outlet) composition, y_i^t, where $i = 1, 2, \ldots, N_R$. In addition, the supply and target compositions, x_j^s and x_j^t, are given for each MSA, where $j = 1, 2, \ldots, N_S$. The flowrate of any lean stream, L_j, is unknown but is bounded by a given maximum available flowrate of that stream, i.e.,

$$L_j \leq L_j^C. \tag{8.1}$$

Figure 8.1 is a schematic illustration of the REAMEN synthesis problem.

As has been previously discussed in the synthesis of physical MENs, several design decisions are to be made:

- Which mass-exchange operations should be used (e.g., absorption, adsorption, etc.)?
- Which MSAs should be selected (e.g., physical/reactive transfer, which solvents, adsorbents, etc.)?
- What is the optimal flowrate of each MSA?
- How should these MSAs be matched with the waste streams (i.e., stream pairings)?
- What is the optimal system configuration (e.g., how should these mass exchangers be arranged?, is there any stream splitting and mixing?, etc.)?

The first step in synthesizing a REAMEN is to establish the conditions for which the reactive mass exchange is thermodynamically feasible. This issue is covered by the next section.

8.2 Corresponding Composition Scales for Reactive Mass Exchange

The fundamentals of reactive mass exchange, design of individual units, chemical equilibrium and kinetics are covered in the literature (e.g., Fogler, 1992; Friedly, 1991; El-Halwagi, 1990; Kohl and Reisenfeld, 1985; Westerterp *et al.*, 1984; Astarita *et al.*, 1983; Smith and Missen, 1982; Espenson, 1981; Levenspiel, 1972; El-Halwagi, 1971). This section presents the salient basics of these systems.

In order to establish the conditions for thermodynamic feasibility of reactive mass exchange, it is necessary to invoke the basic principles of mass transfer with chemical reactions. Consider a lean phase j that contains a set $B_j = \{B_{z,j} \mid z = 1, \ldots, NZ_j\}$ of reactive species (i.e., the set B_j contains NZ_j reactive species, each denoted by $B_{z,j}$, where the index z assumes values from 1 to NZ_j). These species react with the transferable key solute, A, or among themselves via Q_j independent chemical reactions which may be represented by

$$A + \sum_{z=1}^{NZ_j} \nu_{1,z,j} B_{z,j} = 0 \tag{8.2}$$

and

$$\sum_{z=1}^{NZ_j} \nu_{q_j,z,j} B_{z,j} = 0 \quad q_j = 2, 3, \ldots, Q_j, \tag{8.3}$$

where the stoichiometric coefficients $\nu_{q,z,j}$ are negative for products and positive for the reactants. It is worth noting that stoichiometric equations can be mathematically handled as algebraic equations. Therefore, although component A may be involved in more than the first reaction, one can always manipulate the stoichiometric equations algebraically to keep A in the first reaction and eliminate it from the other stoichiometric equations.

Compositions of the different species can be tracked by relating them to the extents of the reactions through the following expression

$$b_{z,j} = b_{z,j}^o - \sum_{q_j=2}^{Q_j} \nu_{q_j,z,j} \xi_{q_j} \quad (z = 1, 2, \ldots, NZ_j), \tag{8.4}$$

where $b_{z,j}$ is the composition of species $B_{z,j}$ in the jth lean phase, $b_{z,j}^o$ is the admissible composition of species $B_{z,j}$ in the jth lean phase and ξ_{q_j} is the extent of the q_jth reaction (or the q_jth reaction coordinate). The extent of reaction is defined for reactions two through Q_j. The reason for not defining it for the first reaction is that the variable u_j plays indirectly the role of the extent of reaction for the first reaction.[1] The admissible compositions may be selected as the lean-phase composition at some particular instant of time, or any other situation which is compatible with stoichiometry and mass-balance bounds.

The equilibrium constant of a reaction is the product of compositions of reactants and products each raised to its stoichiometric coefficient. Hence, for the reaction described by Eq. (8.2) one may write

$$K_{1,j} = \frac{1}{a_j} \prod_{z=1}^{NZ_j} \left(b_{z,j}^o - \sum_{q_j=2}^{Q_j} \nu_{q_j,z,j} \xi_{q_j} \right)^{\nu_{1,z,j}} ,$$

i.e.,

$$a_j = \frac{1}{K_{1,j}} \prod_{z=1}^{NZ_j} \left(b_{z,j}^o - \sum_{q_j=2}^{Q_j} \nu_{q_j,z,j} \xi_{q_j} \right)^{\nu_{1,z,j}} \tag{8.5}$$

and for the reactions given by Eq. (8.3):

$$K_{q_j,j} = \prod_{z=1}^{NZ_j} \left(b_{z,j}^o - \sum_{q_j=2}^{Q_j} \nu_{q_j,z,j} \xi_{q_j} \right)^{\nu_{q_j,z,j}} , \quad q_j = 2, 3, \ldots, Q_j, \tag{8.6}$$

where $K_{q_j,j}$ is the equilibrium constant for the q_jth reaction and a_j is the composition of the physically dissolved A in lean phase j.

It is now useful to recall the concepts of *molarity and fractional saturation* (Astarita *et al.*, 1983). The molarity, m_j, of a reactive MSA is the total equivalent concentration of species that may react with component A. On the other hand, the fractional saturation, u_j, is a variable that represents the degree of saturation of chemically combined A in the jth lean phase. Therefore, $u_j m_j$ is the total concentration of chemically combined A in the jth MSA. Hence, the total concentration of A in MSA j can be expressed as

$$x_j = a_j + u_j m_j, \quad j \in S, \tag{8.7}$$

where the physically-dissolved concentration of A, a_j, equilibrates with the rich-phase composition through a distribution function F_j, i.e.,

$$y_i^* = F_j(a_j), \quad j \in S. \tag{8.8}$$

Equations (8.4)–(8.8) represent a complete mathematical description of the chemical equilibrium between a rich phase and the jth MSA. The simultaneous solution

---

[1] For this reason, whenever there is a single reaction taking place in the jth MSA, no extent of reaction is defined. Instead the fractional saturation, u_j, is employed.

of these nonlinear equations (for instance by using the software LINGO) yields the equilibrium compositions of both phase in the form

$$y_i^* = f_j(x_j^*) \quad j \in S. \tag{8.9}$$

For any mass-exchange operation to be thermodynamically feasible, the following conditions must be satisfied:

$$x_j < x_j^*, \quad j \in S, \tag{8.10a}$$

and/or

$$y_i > y_i^*, \quad i \in R, \tag{8.10b}$$

i.e.,

$$y = f(x_j + \varepsilon_j^s), \quad j \in S, \tag{8.11a}$$

where

$$y = y_i - \varepsilon_i^R, \quad i \in R, \tag{8.11b}$$

where ε_i^R and ε_j^S are positive quantities called, respectively, the rich and the lean minimum allowable composition differences. The parameters ε_i^R and ε_j^S are optimizable quantities that can be used for trading off capital versus operating costs (see Chapters Two and Three). Equations (8.11a) and (8.11b) provides a correspondence among the rich and the lean composition scales for which mass exchange is practically feasible. This is the reactive equivalent to Eq. (3.5) used for establishing the corresponding composition scales for physical MENs.

Example 8.1: Absorption of H_2S in Aqueous Sodium Hydroxide

Consider an aqueous caustic soda solution whose molarity $m_1 = 5.0$ kmol/m^3 (20 wt.% NaOH). This solution is to be used in absorbing H_2S from a gaseous waste. The operating range of interest is $0.0 \leq x_1 \ (kmol/m^3) \leq 5.0$. Derive an equilibrium relation for this chemical absorption over the operating range of interest.

Solution

The absorption of H_2S in this solution is accompanied by the following chemical reactions (Astarita and Gioia, 1964):

$$H_2S + NaOH = NaHS + H_2O \tag{8.12}$$

$$H_2S + 2NaOH = Na_2S + 2H_2O. \tag{8.13}$$

As has been described earlier, the stoichiometric reactions should be manipulated algebraically to retain the transferable species (H_2S) only in the first equation. Therefore, H_2S can be eliminated from Eq. (8.13) by subtracting (8.12) from (8.13) to get

$$NaHS + NaOH = Na_2S + H_2O. \qquad (8.14)$$

Equations (8.12) and (8.14) can be written in ionic terms as follows:

$$H_2S + OH^- = HS^- + H_2O \qquad (8.15)$$

$$HS^- + OH^- = S^{2-} + H_2O. \qquad (8.16)$$

Let us denote the three ionic species as follows:

$$OH^- = B_{1,1}$$

$$HS^- = B_{2,1}$$

$$S^{2-} = B_{3,1},$$

with aqueous-phase concentrations referred to as $b_{1,1}$, $b_{2,1}$ and $b_{3,1}$, respectively. Also, let us denote the composition of the physically dissolved H_2S in the aqueous solution as a_1.

For cases when the concentration of water remains almost constant with respect to the other species, one can define the following reaction equilibrium constants for Eqs. (8.15) and (8.16), respectively:

$$K_{1,1} = \frac{b_{2,1}}{a_1 \cdot b_{1,1}} \qquad (8.17)$$

and

$$K_{2,1} = \frac{b_{3,1}}{b_{2,1} \cdot b_{1,1}}, \qquad (8.18)$$

where $K_{1,1} = 9.0 \times 10^6$ m^3/kmol and $K_{2,1} = 0.12$ m^3/kmol.
The distribution coefficient for the physically dissolved H_2S is given by

$$\frac{y_1}{a_1} = 0.368, \qquad (8.19)$$

where y_l is the composition of H_2S in the gaseous stream.

It is useful to relate the molarity of the aqueous caustic soda ($m_1 = 5.0$ kmol NaOH/m^3) to the other reactive species. Once the reactions start, the composition of NaOH will decrease. However, it is possible to relate the molarity of the solution to the concentration of the reactive species at any reaction coordinate. Suppose that after a certain extent of reaction (8.15) and (8.16) an analyzer is placed in the solution to measure the compositions of OH^-, HS^- and S^{2-} with the measured concentrations denoted by $b_{1,1}$, $b_{2,1}$ and $b_{3,1}$, respectively. These measured concentrations are related to the molarity as follows. According to Eq. (8.15), $b_{2,1}$ kmoles of OH^- must have reacted to yield $b_{2,1}$ kmoles of HS^-. Similarly, according to Eq. (8.16), $b_{3,1}$ kmoles of OH^- must have reacted with $b_{3,1}$ kmoles of HS^- to yield $b_{3,1}$ kmoles of S^{2-}. But the $b_{3,1}$ kmoles of HS^- must have resulted from the reaction of $b_{3,1}$ kmoles of OH^- (according to Eq. (8.15)). Hence, $2b_{3,1}$ kmoles of OH^- are consumed in producing $b_{3,1}$ kmoles of S^{2-}. Therefore, the molarity of the aqueous caustic soda can be related to the concentrations of the reactive ions as follows:

$$5 = b_{1,1} + b_{2,1} + 2b_{3,1}. \qquad (8.20)$$

There are two forms of reacted H_2S in the solution: HS^- and S^{2-}. By recalling that $u_1 m_1$ is the total concentration of chemically combined H_2S in the aqueous caustic soda and conducting an atomic balance on S over Eqs. (8.15) and (8.16), we get

$$5u_1 = b_{2,1} + b_{3,1}. \qquad (8.21)$$

As discussed earlier, the admissible compositions may be selected as the lean-phase composition at some particular instant of time, or any other situation which is compatible with stoichiometry and mass-balance bounds such as Eqs. (8.20) and (8.21). Let us arbitrarily select the admissible composition of S^{2-} to be zero, i.e.,

$$b_{3,1}^0 = 0. \qquad (8.22)$$

This selection automatically fixes the corresponding values of $b_{1,1}^0$ *and* $b_{2,1}^0$. According to Eqs. (8.20) and (8.22), we get

$$b_{2,1}^0 = 5u_1. \qquad (8.23)$$

Similarly, according to Eqs. (8.21)–(8.23), we obtain:

$$b_{1,1}^0 = 5(1 - u_1). \qquad (8.24)$$

As mentioned earlier, the extent of reaction is defined for all reactions except the first one. Hence, we define ξ_2 as the extent of reaction for Eq. (8.16). Equation (8.4) can now be used to describe the compositions of the reactive species as a function of ξ_2 and the admissible compositions, i.e.,

$$b_{1,1} = 5(1 - u_1) - \xi_2 \qquad (8.25)$$

$$b_{2,1} = 5u_1 - \xi_2 \qquad (8.26)$$

$$b_{3,1} = \xi_2. \qquad (8.27)$$

Substituting from Eqs. (8.19), (8.25), (8.26) and (8.27) into Eq. (8.17), we get

$$9.0 \times 10^6 = \frac{5u_1 - \xi_2}{\dfrac{y_1}{0.368}[5(1 - u_1) - \xi_2]}$$

or

$$y_1 = 4.09 \times 10^{-8} \frac{5u_1 - \xi_2}{[5(1 - u_1) - \xi_2]}. \tag{8.28}$$

Substituting from Eqs. (8.25)–(8.27) into Eq. (8.18), we obtain

$$0.12 = \frac{\xi_2}{[5u_1 - \xi_2][5(1 - u_1) - \xi_2]},$$

which can be rearranged to

$$\xi_2^2 - 13.33\,\xi_2 + 25u_1(1 - u_1) = 0. \tag{8.29}$$

According to Eq. (8.7), the total concentration of H_2S in the aqueous caustic soda (physically dissolved and chemically reacted) can be expressed as follows

$$x_1 = a_1 + 5u_1. \tag{8.30a}$$

In this system, the physically dissolved H_2S is much less than the chemically combined,[2] i.e.,

$$a_1 \ll 5u_1, \tag{8.30b}$$

which simplifies Eq. (8.30a) to

$$x_1 = 5u_1. \tag{8.30c}$$

Equations (8.28), (8.29), and (8.30c) can be used to develop an expression for reactive mass exchange of H_2S. First, a value of fractional saturation is selected (where $x_1^S/5 \leq u_1 \leq x_1^t/5$). Then, Eq. (8.30c) is used to calculate the corresponding x_1. Next, Eq. (8.29) is solved to determine the value of ξ_2. Finally, Eq. (8.28) is solved to evaluate y_1. The pair (y_1, x_1) are in equilibrium. The same procedure is repeated for several values of u_1 (between $x_1^S/5$ and $x_1^t/5$) to yield pairs of (y_1, x_1) which are in equilibrium. Nonlinear regression can be employed to derive an equilibrium expression for these pairs.

To illustrate the above-mentioned procedure, let us start with a fractional saturation of $u_1 = 0.1$. According to Eq. (8.30c), $x_1 = 0.5\,\text{kmol/m}^3$. By substituting for $u_1 = 0.1$ in Eq. (8.29), we obtain:

$$\xi_2^2 - 13.33\xi_2 + 2.25 = 0,$$

---

[2]This assumption can be numerically verified by comparing values of a_j with $u_j m_j$ after the equilibrium equation is generated.

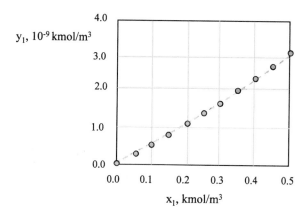

Figure 8.2 Equilibrium data for the example of absorbing H_2S in caustic soda.

which can be solved to get

$$\xi_2 = \frac{13.33 - \sqrt{13.33 \times 13.33 - 4 \times 2.25}}{2}$$

$$= 0.171 \, kmol/m^3 \, (\text{the other root is rejected}) \tag{8.31}$$

The values of u_1 and ξ_2 are plugged into Eq. (8.28) to get $y_1 = 3.1 \times 10^{-9} \, kmol/m^3$. Hence, the pair $(3.1 \times 10^{-9}, 0.5)$ are in equilibrium.[3] The same procedure is repeated for values of u_1 between 0.0 and 0.1. The results are plotted as shown in Fig. 8.2. The plotted data are slightly convex and can be fitted to the following quadratic function (all compositions in $kmol/m^3$):

$$y_1 = 2.364 \times 10^{-9} x_1^2 + 5.022 \times 10^{-9} x_1 \tag{8.32}$$

with a correlation coefficient, r^2, of 0.9999.

8.3 Synthesis Approach

Now that a procedure for establishing the corresponding composition scales for the rich-lean pairs of stream has been outlined, it is possible to develop the CID. The CID is constructed in a manner similar to that described in Chapter Five. However, it should be noted that the conversion among the corresponding composition scales may be more laborious due to the nonlinearity of equilibrium relations. Furthermore, a lean scale, x_j, represents all forms (physically dissolved and chemically combined) of the pollutant. First, a composition scale, y, for component A in

---

[3]We can now test the validity of Eq. (8.30b). According to Eq. (8.19), $a_1 = 8.45 \times 10^{-9} \, kmol/m^3$ which is indeed much less than $u_1 m_1 = 0.5$.

any rich stream is created. This scale is in a one-to-one correspondence with any composition scale of component A in the ith rich stream, y_i, via Eq. (8.11b). Then, Eq. (8.11a) is used to generate N_S composition scales for component A in the lean streams. Next, each stream is represented by an arrow whose tail and head correspond to the supply and target compositions, respectively, of the stream. The partitions corresponding to these heads and tails establish the composition intervals. Similar to the CID for physical MENs, within any composition interval, it is thermodynamically feasible to transfer component A from the rich streams to the lean streams. Also, according to the second law of thermodynamics it is spontaneously possible to transfer component A from the rich streams in a given composition interval to any lean stream within a lower composition interval.

It is worth noting that the foregoing composition partitioning procedure ensures thermodynamic feasibility only when all the equilibrium relations described by Eq. (8.9) are convex. In this case by merely satisfying Eq. (8.11) at both ends of a composition interval, Eqs. (8.10) are automatically satisfied throughout that interval (Fig. 8.3a). On the other hand, when at least one of the equilibrium relations expressed by Eq. (8.9) is non-convex, the satisfaction of Eq. (8.11) at both ends of an interval does not necessarily imply the realization of inequalities (Eqs. 8.10) throughout that interval (Fig. 8.3b). In such case, additional composition partitioning is needed. This can be achieved by discretizing the non-convex portions of the equilibrium curves through linear

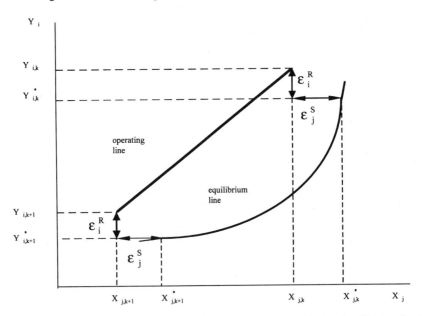

Figure 8.3a A reactive mass exchanger with convex equilibrium (El-Halwagi and Srinivas, Synthesis of reactive mass-exchange networks, *Chem. Eng. Sci.*, **47**(8), p. 2115, Copyright © 1992, with kind permission from Elsevier Science Ltd., The Boulevard, Langford Lane, Kidlington 0X5 1GB, UK).

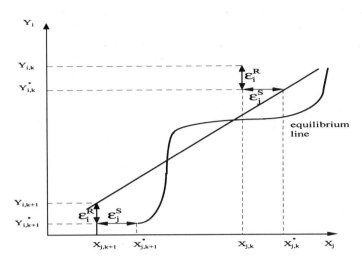

Figure 8.3b A reactive mass exchanger with non-convex equilibrium (El-Halwagi and Srinivas, Synthesis of reactive mass-exchange networks, *Chem. Eng. Sci.*, **47**(8), p. 2115, Copyright © 1992, with kind permission from Elsevier Science Ltd., The Boulevard, Langford Lane, Kidlington 0X5 1GB, UK).

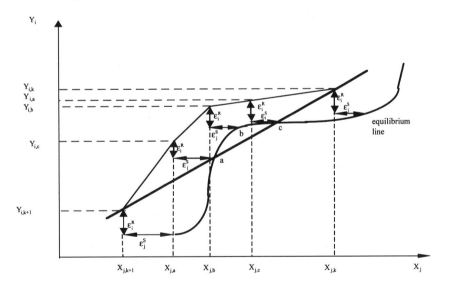

Figure 8.3c A reactive mass exchanger with discretization of non-convex equilibrium (El-Halwagi and Srinivas, Synthesis of reactive mass-exchange networks, *Chem. Eng. Sci.*, **47**(8), p. 2115, Copyright © 1992, with kind permission from Elsevier Science Ltd., The Boulevard, Langford Lane, Kidlington 0X5 1GB, UK).

overestimators.[4] Clearly, the number of these linear segments depends upon the desired degree of accuracy. For instance, Fig. 8.3b may be convexified by introducing two linear parts at the concavity inflection point (Fig. 8.3c). Consequently, additional intervals will be created within the CID by partitioning the composition scales at the locations corresponding to points a, b and c. The end result of this partitioning procedure is that the entire composition range is divided into n_{int} composition intervals, with $k = 1$ being the highest and $k = n_{int}$ being the lowest.

Having established a one-to-one thermodynamically-feasible correspondence among all the composition scales, we can now solve the REAMEN problem via a transshipment formulation similar to that described in Chapter Five for the synthesis of physical MENs.

Example 8.2: Removal of H₂S from a Kraft Pulping Process

Kraft pulping is a common process in the paper industry. Figure 8.4 shows a simplified flowsheet of the process. In this process, wood chips are reacted (cooked) with white liquor in a digester. White liquor (which contains primarily NaOH, Na_2S, Na_2CO_3 and water) is employed to dissolve lignin from the wood chips. The cooked pulp and liquor are passed to a blow tank where the pulp is separated from the spent liquor "weak black liquor" which is fed to a recovery system for

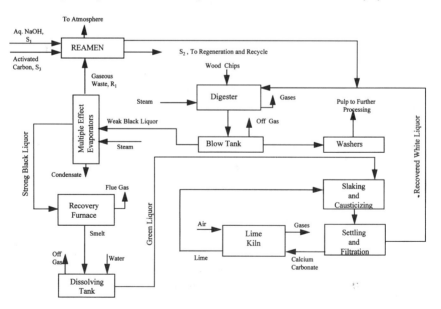

Figure 8.4 Kraft pulping process.

---

[4]A more rigorous method of tackling nonconvex equilibrium without discretization has been developed by Srinivas and El-Halwagi (1994) and is beyond the scope of this book.

Table 8.1
Data for the Gaseous Emission of the Kraft Pulping Process

Stream	Description	Flowrate G_i, m³/s	Supply composition $(10^{-10}$ kmol/m³$)$ y_i^s	Target composition $(10^{-10}$ kmol/m³$)$ y_i^t
R_1	Gaseous waste from evaporators	16.2	1,600	3.0

conversion to white liquor. The first step in recovery is concentration of the weak liquor via multiple effect evaporators. The concentrated solution is sprayed in a furnace. The smelt from the furnace is dissolved in water to form green liquor which is reacted with lime (CaO) to produce white liquor and calcium carbonate "mud". The recovered white liquor is mixed with make-up materials and recycled to the digester. The calcium carbonate mud is thermally decomposed in a kiln to produce lime which is used in the causticizing reaction.

There are several gaseous wastes emitted from the process (see Dunn and El-Halwagi, 1993, and Homework Problem 8.5). In this example, we focus on the gaseous waste leaving the multiple effect evaporators, R_1, whose primary pollutant is H₂S. Stream data for this waste stream are given in Table 8.1. A rich-phase minimum allowable composition difference, ε_i^R, of 1.5×10^{-10} kmol/m³ is used.

A process lean stream and an external MSA are considered for removing H₂S. The process lean stream, S_1, is a caustic soda solution which can be used as a solvent for the reactive separation of H₂S. An added bonus for using the process MSA is the conversion of a portion of the absorbed H₂S into Na₂S, which is needed for white-liquor makeup. In other words, H₂S "the pollutant" is converted into a valuable chemical which is needed in the process. The external MSA, S_2, is a polymeric adsorbent. The data for the candidate MSAs are given in Table 8.2. The equilibrium

Table 8.2
Data for the MSAs of the Kraft Pulping Example

Stream	Upper bound on flowrate L_j^C m³/s	Supply composition (kmol/m³) x_j^s	Target composition (kmol/m³) x_j^t	ε_j^s (kmol/m³)	C_j $/m³ MSA
S_1	0.01	0.000	0.500	0.10	–
S_2	∞	0.010	0.025	0.01	2,900

data for the transfer of H_2S from the waste stream to the adsorbent is given by

$$y_1 = 2 \times 10^{-9} x_2, \tag{8.33}$$

where y_1 and x_2 are given in $kmol/m^3$.

For the given data, determine the minimum operating cost of the REAMEN and construct a network with the minimum number of exchangers.

Solution

The following expression for the equilibrium data of H_2S in caustic soda has been derived in Eq. (8.32) of Example 8.1 (all compositions are in $kmol/m^3$):

$$y_1 = 2.364 \times 10^{-9} + 5.022 \times 10^{-9} x_1. \tag{8.32}$$

By invoking Eq. (8.11), we get the following equation for the practical-feasibility curve:

$$y_1 - 1.5 \times 10^{-10} = 2.364 \times 10^{-9}(x_1 + 0.1)^2$$
$$+ 5.022 \times 10^{-9}(x_1 + 0.1). \tag{8.34}$$

Similarly, for activated carbon

$$y_1 - 1.5 \times 10^{-10} = 2 \times 10^{-9}(x_2 + 0.01). \tag{8.35}$$

The CID for the problem is shown in Fig. 8.5. The table of exchangeable loads for the waste streams and the MSAs are shown by Tables 8.3 and 8.4, respectively.

According to the two-stage targeting procedure, we first minimize the annual operating cost of the MSAs (given by $3600 \times 8760 \times 2100 \times L_2 = 6.62256 \times 10^{10} L_2$ when we assume 8760 operating hours per year). By applying the linear-

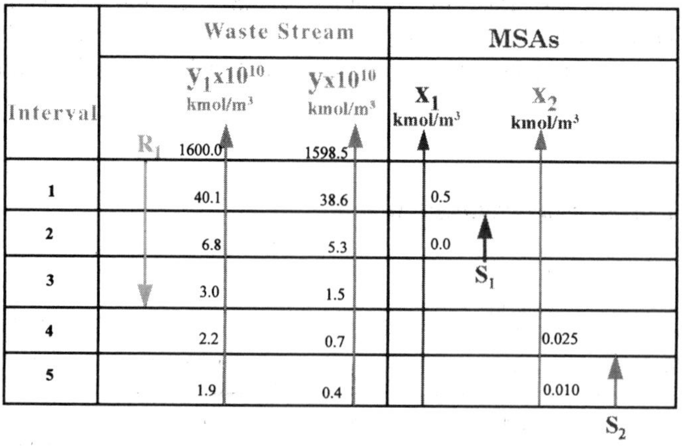

| Interval | Waste Stream | | MSAs | |
	$y_1 \times 10^{10}$ $kmol/m^3$	$y \times 10^{10}$ $kmol/m^3$	X_1 $kmol/m^3$	X_2 $kmol/m^3$
R_1	1600.0	1598.5		
1	40.1	38.6	0.5	
2	6.8	5.3	0.0	
3	3.0	1.5	S_1	
4	2.2	0.7		0.025
5	1.9	0.4		0.010
				S_2

Figure 8.5 CID for Kraft pulping example.

Table 8.3
The TEL for the Gaseous Waste

Interval	Load of R$_1$ (10^{-10} kmol H$_2$S/s)
1	25,270
2	559
3	62
4	0
5	0

programming formulation described in Chapter Six (P6.1), one can write the following optimization program:

$$\min 6.62256 \times 10^{10} L_2, \qquad \text{(P8.1)}$$

subject to

$$\delta_1 = 25,270 \times 10^{-10}$$

$$\delta_2 - \delta_1 + 0.5L_1 = 559 \times 10^{-10}$$

$$\delta_3 - \delta_2 = 62 \times 10^{-10}$$

$$\delta_4 - \delta_3 = 0.0$$

$$-\delta_4 + 0.015L_2 = 0.0$$

$$\delta_k \geq 0 \quad k = 1, 2, 3, 4$$

$$L_j \geq 0 \quad j = 1, 2$$

$$L_1 \leq 0.01 \times 10^{-10}.$$

Table 8.4
The TEL for the Lean Streams

Interval	Capacity of lean streams per m^3 of MSA (kmol H$_2$S/m^3 MSA)	
	S$_1$	S$_2$
1	–	–
2	0.5	–
3	–	–
4	–	–
5	–	0.015

It is worth pointing out that the wide range of coefficients may cause computational problems for the optimization software. This is commonly referred to as the "scaling" problem. One way of circumventing this problem is to define scaled flowrates of MSAs in units of 10^{-10} m^3/s and scaled residual loads in units of 10^{-10} kmol/s, i.e., let

$$L_j^{scaled} = 10^{10} L_j, \quad j = 1, 2 \tag{8.36}$$

$$\delta_k^{scaled} = 10^{10} \delta_k, \quad k = 1, 2, 3, 4 \tag{8.37}$$

With the new units, the scaled program becomes:

$$\min 6.62256 \, L_2^{scaled} \tag{P8.2}$$

subject to

$$\delta_1^{scaled} = 25,270$$

$$\delta_2^{scaled} - \delta_1^{scaled} + 0.5 L_1^{scaled} = 559$$

$$\delta_3^{scaled} - \delta_2^{scaled} = 62$$

$$\delta_4^{scaled} - \delta_3^{scaled} = 0.0$$

$$-\delta_4^{scaled} + 0.015 L_2^{scaled} = 0.0$$

$$\delta_k^{scaled} \geq 0 \quad k = 1, 2, 3, 4$$

$$L_j^{scaled} \geq 0 \quad j = 1, 2$$

$$L_1^{scaled} \leq 0.01$$

In terms of LINGO input, the scaled program can be written as follows (with the S in each variable indicating that it is a scaled variable):

```
model:
min = 6.62256*LS2;
deltaS1 = 25270;
deltaS2 - deltaS1 + 0.5*LS1 = 559;
deltaS3 - deltaS2 = 62;
deltaS4 - deltaS3 = 0.0;
- deltaS4 + 0.015*LS2 = 0.0;
deltaS1 >= 0.0;
deltaS2 >= 0.0;
deltaS3 >= 0.0;
deltaS4 >= 0.0;
LS1 >= 0.0;
LS2 >= 0.0;
end
```

The solution report from LINGO gives the following results:

```
Objective value:    27373
    Variable        Value
      LS2           4133.333
    DELTAS1         25270.00
    DELTAS2         0.0000000E+00
      LS1           51658.00
    DELTAS3         62.00000
    DELTAS4         62.00000
```

Therefore, the minimum operating cost is approximately \$27,000/year and the pinch location is between the second and third composition intervals. The REAMEN involves two exchangers; one above the pinch matching R_1 with S_1 and one below the pinch matching R_1 with S_2.

Problems

8.1 Derive an equilibrium expression for the reactive absorption of H_2S in diethanolamine "DEA". The molarity of DEA is 2 kmol/m^3. The following reaction takes place:

$$H_2S + (C_2H_5)_2NH = HS^- + (C_2H_5)_2NH_2^+,$$

whose equilibrium constant is given by (Lal *et al.*, 1985):

$$K = 234.48 = \frac{[HS^-][(C_2H_5)_2[NH_2^+]}{[H_2S][(C_2H_5)_2NH]} \tag{8.38}$$

with all concentrations in kmol/m^3. The physical distribution coefficient is:

$$0.363 = \frac{\text{Composition of } H_2S \text{ in gas (kmol/m}^3)}{\text{Composition of physically-dissolved } H_2S \text{ in DEA (kmol/m}^3)} \tag{8.39}$$

8.2 Develop equilibrium equations for the reactive absorption of CO_2 into:

(a) Aqueous potassium carbonate
(b) Monoethanolamine

(**Hint:** See pp. 68–79 of Astarita *et al.*, 1983.)

8.3 Coal may be catalytically hydrogenated to yield liquid transportation fuels. A simplified process flow diagram of a coal-liquefaction process is shown in Fig. 8.6. Coal is mixed with organic solvents to form a slurry which is reacted with hydrogen. The reaction products are fractionated into several transportation fuels. Hydrogen sulfide is among the primary gaseous pollutants of the process (Warren *et al.*, 1995). Hence, it is desired to design a cost-effective H_2S recovery system.

Two major sources of H_2S emissions from the process are the acid gas stream evolving from hydrogen manufacture, R_1, and the gaseous waste emitted from the separation

Table 8.5
Data for the Waste Streams of the Coal-Liquefaction Problem (All Compositions Are in kmol/m³)

Stream	Flowrate L_j^C m³/hr	Supply composition y_j^s kmol/m³	Target composition y_j^t kmol/m³
R_1	121.1	3.98×10^{-4}	2.1×10^{-7}
R_2	28.9	71.6×10^{-4}	2.1×10^{-7}

section of the process, R_2, as shown in Fig. 8.6. Stream data for these acid gas streams are summarized in Table 8.5.

Six potential MSAs should be simultaneously screened. These include absorption in water, S_1, adsorption onto activated carbon, S_2, absorption in chilled methanol, S_3, and the use of the following reactive solvents; diethanolamine (DEA), S_4, hot potassium carbonate, S_5, and diisopropanolamine (DIPA), S_6.

Equilibrium relations governing the transfer of hydrogen sulfide from the gaseous waste streams to the various separating agents can be approximated over the range of operating

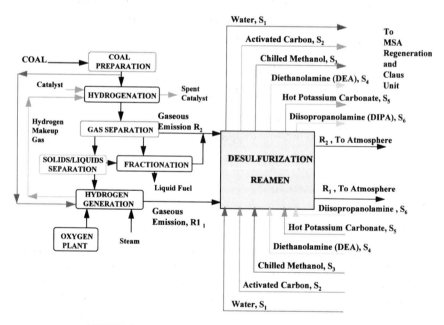

A SIMPLIFIED DIAGRAM OF A COAL LIQUEFACTION PROCESS

Figure 8.6 Coal-liquefaction process (Warren et al., 1995. Courtesy of ASCE).

Table 8.6
Data for the MSAs of the Coal-Liquefaction Problem
(All Compositions Are in kmol/m³)

Stream	L_j^C m³/s	Supply composition x_j^s 10^{-6} kmol/m³	Target composition x_j^t kmol/m³
S_1	∞	1000	0.0150
S_2	∞	2.00	0.4773
S_3	∞	2.15	0.2652
S_4	∞	3.40	0.1310
S_5	∞	3.20	0.0412
S_6	∞	1.30	0.7160

compositions (kmol/m³) as follows:

$$y = 0.398x_1 \tag{8.40}$$

$$y = 0.015x_2 \tag{8.41}$$

$$y = 0.027x_3 \tag{8.42}$$

$$y = 0.079x_4^{2.6} \tag{8.43}$$

$$y = 0.013x_5 \tag{8.44}$$

$$y = 0.010x_6 \tag{8.45}$$

The unit operating costs of water, S_1, activated carbon, S_2, chilled methanol, S_3, diethanolamine, S_4, hot potassium carbonate, S_5, and diisopropanolamine, S_6, are 0.001, 8.34, 2.46, 5.94, 3.97, and 4.82 in \$/m³, respectively. These costs include the cost of regeneration and make-up. Stream data for the MSA's are given in Table 8.6. The values of ε_i^R and ε_j^S are taken to be 0.0 and 2×10^{-7} kmol/m³, respectively.

Synthesize a REAMEN which features the minimum number of units that realize the minimum operating cost.

8.4 Most of the world's rayon is produced through the viscose process (El-Halwagi and Srinivas, 1992). Figure 8.7 provides a schematic representation of the process. In this process cellulose pulp is treated with caustic soda, then reacted with carbon disulfide to produce cellulose xanthate. This compound is dissolved in dilute caustic soda to give a viscose syrup, which is fed to vacuum-flash boiling deaerator to remove air. The gaseous stream leaving the deaerator, R_1, should be treated for the removal of H_2S prior to its atmospheric discharge. In spinning, a viscose solution is extruded through fine holes submerged in an acid bath to produce the rayon fibers. The acid-bath solution contains sulfuric acid, which neutralizes caustic soda and decomposes xanthate and various sulfur-containing species, thus producing H_2S as the major hazardous compound in the exhaust gas stream, R_2.

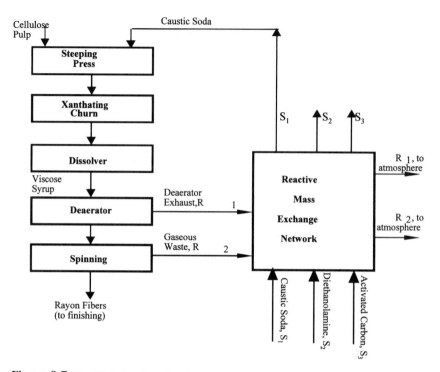

Figure 8.7 Simplified flowsheet for viscose rayon production (from El-Halwagi and Srinivas, Synthesis of reactive mass-exchange networks, *Chem. Eng. Sci.*, **47**(8), p. 2116, Copyright © 1992, with kind permission from Elsevier Science Ltd., The Boulevard, Langford Lane, Kidlington 0X5 1GB, UK).

It is desired to synthesize a REAMEN for treating the gaseous wastes (R_1 and R_2) of a viscose rayon plant. Three MSAs are available to select from. These MSAs are caustic soda, S_1, (a process stream already existing in the plant with $m_1 = 5.0$ kmol/m^3), diethanolamine, S_2, (with $m_2 = 2.0$ kmol/m^3) and activated carbon, S_3. The unit costs for S_2 and S_3 including stream makeup and subsequent regeneration are 64.9 $/m^3 and 169.4 $/m^3, respectively. Stream data are given in Table 8.7.

The chemical absorption of H_2S into caustic-soda solution (Astarita and Gioia, 1964) involves the following two reactions:

$$H_2S + OH^- \rightarrow HS^- + H_2O \qquad (8.46)$$

$$HS^- + OH^- \rightarrow S^{2-} + H_2O. \qquad (8.47)$$

Since the concentration of water remains approximately constant, one can define the following two reaction equilibrium constants

$$K_{1.1} = \frac{[HS^-]}{[H_2S][OH^-]} \qquad (8.48)$$

Table 8.7
Stream Data for the Viscose-Rayon Example

		Rich streams			Lean streams		
Stream	G_i	y_i^s (kmol/m^3)	y_i^t (kmol/m^3)	Stream	L_j^C (m^3/s)	x_j^s (kmol/m^3)	x_j^t (kmol/m^3)
R_1	0.87	1.3×10^{-5}	2.2×10^{-7}	S_1	2.0×10^{-4}	0.0	0.1
R_2	0.10	0.9×10^{-5}	2.2×10^{-7}	S_2	∞	2.0×10^{-6}	1.0×10^{-3}
				S_3	∞	1.0×10^{-6}	3.0×10^{-6}

and

$$K_{2,1} = \frac{[S^{2-}]}{[HS^-][OH^-]} \tag{8.49}$$

where $K_{1,1} = 9.0 \times 10^6$ m^3/kmol and $K_{2,1} = 0.12$ m^3/kmol. The distribution coefficient for the physically dissolved portion of H_2S between a rich phase and caustic-soda solution is given by

$$y_i/a_1 = 0.368. \tag{8.50}$$

The overall reaction of hydrogen sulfide with diethanol amine (Lal *et al.*, 1985) is given by

$$H_2S + (C_2H_4OH)_2NH \Longleftrightarrow HS^- + (C_2H_4OH)_2NH_2^+, \tag{8.51}$$

for which the equilibrium constant is given by

$$K_{1,2} = 234.48 = \frac{[HS^-][R_2NH_2^+]}{[H_2S][R_2NH]} \tag{8.52}$$

and the physical distribution coefficient is given by (Kent and Eisenberg, 1976):

$$y_i/a_2 = 0.363. \tag{8.53}$$

The adsorption isotherm for H_2S on activated carbon is represented by (Valenzuela and Myers, 1989)

$$y_i/a_3 = 0.015. \tag{8.54}$$

The following values for the minimum allowable composition differences are selected:

$$\varepsilon_1^S = 3.50, \varepsilon_2^S = 0.014, \varepsilon_3^S = 10^{-6},$$

$$\varepsilon_1^R = 10^{-7}, \varepsilon_2^R = 1.5 \times 10^{-7} \text{kmol/m}^3$$

8.5 Consider the Kraft pulping process shown in Fig. 8.8 (Dunn and El-Halwagi, 1993). The first step in the process is digestion in which wood chips, containing primarily lignin, cellulose, and hemicellulose, are "cooked" in white liquor (NaOH, Na$_2$S, Na$_2$CO$_3$ and

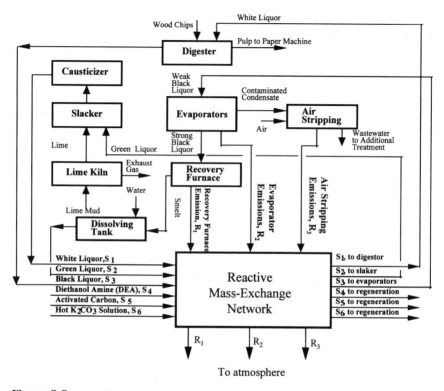

Figure 8.8 A simplified flowsheet of the Kraft pulping process (Dunn and El-Halwagi, 1993).

water) to solubilize the lignin. The off gases leaving the digester contain substantial quantities of H_2S. The dissolved lignin leaves the digester in a spent solution referred to as the "weak black liquor". This liquor is processed through a set of multiple-effect evaporators designed to increase the solid content of this stream from approximately 15% to approximately 65%. At the higher concentration, this stream is referred to as strong black liquor. The contaminated condensate removed through the evaporation process can be processed through an air stripper to transfer sulfur compounds (primarily H_2S) to an air stream prior to further treatment and discharge of the condensate stream. The strong black liquor is burned in a furnace to supply energy for the pulping processes and to allow the recovery of chemicals needed for subsequent pulp production. The burning of black liquor yields an inorganic smelt (Na_2CO_3 and Na_2S) that is dissolved in water to produce green liquor (NaOH, Na_2S, Na_2CO_3 and water), which is reacted with quick lime (CaO) to convert the Na_2CO_3 into NaOH. The conversion of the Na_2CO_3 into NaOH is referred to as the causticizing reaction and involves two reactions. The first reaction is the conversion of calcium oxide to calcium hydroxide in the presence of water in an agitated slaker. The calcium hydroxide subsequently reacts with Na_2CO_3 to form NaOH and a calcium carbonate precipitant. The calcium carbonate is then heated in the lime kiln to regenerate the calcium oxide and release

Table 8.8
Data for the Gaseous Emission of the Kraft Pulping Process

Stream	Description	Flowrate G_i, m^3/s	Supply composition $(10^{-5}$ kmol/m$^3)$ y_i^s	Target composition $(10^{-7}$ kmol/m$^3)$ y_i^t
R_1	Gaseous waste from recovery furnace	117.00	3.08	2.1
R_2	Gaseous waste from evaporator	0.43	8.2	2.1
R_3	Gaseous waste from air stripper	465.80	1.19	2.1

carbon dioxide. These reactions result in the formation of the original white liquor for reuse in the digesting process.

Three major sources in the kraft process are responsible for the majority of the H_2S emissions. These involve the gaseous waste streams leaving the recovery furnace, the evaporator and the air stripper, respectively denoted by R_1, R_2 and R_3. Stream data for the gaseous wastes are summarized in Table 8.8. Several candidate MSAs are screened. These include three process MSAs and three external MSAs. The process MSAs are the white, the green and the black liquors (referred to as S_1, S_2 and S_3, respectively). The external MSAs include diethanolamine (DEA), S_4, activated carbon, S_5, and 30 wt% hot potassium carbonate solution, S_6. Stream data for the MSAs is summarized in Table 8.9. Synthesize a MOC REAMEN that can accomplish the desulfurization task for the three waste streams.

Table 8.9
Data for the MSAs of the Kraft Pulping Example

Stream	Description	Upper bound on flowrate L_j^C m^3/s	Supply composition (kmol/m^3) x_j^s	Target composition (kmol/m^3) x_j^t
S_1	White liquor	0.040	0.320	3.100
S_2	Green liquor	0.049	0.290	1.290
S_3	Black liquor	0.100	0.020	0.100
S_4	DEA	∞	2×10^{-6}	0.020
S_5	Activated carbon	∞	1×10^{-6}	0.002
S_6	Hot Potassium carbonate	∞	0.003	0.280

Symbols

a_j	physically dissolved concentration of A in lean stream j, kmol/m³
A	key transferable component
$b_{z,j}$	composition of reactive species $B_{z,j}$ in lean phase j, kmol/m³
$b_{z,j}^o$	admissible composition of reactive species $B_{z,j}$ in lean phase j, kmol/m³
B_j	set of reactive species in lean stream j
$B_{z,j}$	index of reactive species in lean stream j
c_j	unit cost of the jth MSA j, ($/kg)
f_j	equilibrium distribution function between rich phase composition and total content of A in lean phase j, defined in Eq. (8.9)
F_j	equilibrium distribution function between rich phase composition and physically dissolved A in lean phase j, defined in Eq. (8.8)
G_i	flow rate of rich stream i, kmol/s
i	index for rich streams
j	index for lean streams
k	index for composition intervals
$K_{q_j,j}$	equilibrium constant for the q_jth reaction in lean phase j
L_j	flow rate of lean stream j, kmol/s
L_j^c	upper bound on flow rate of lean stream j, kmol/s index for subnetworks
m_j	molarity of lean stream j, kmol/m³
n_p	number of mass-exchange pinch points in the problem
N_R	number of rich streams
N_S	number of lean streams
NZ_j	number of reactive species in lean stream j
q_j	index for the independent reactions in lean stream j
Q_j	number of independent reactions in lean stream j
r^2	correlation coefficient
R	set of rich streams
R_i	the ith rich stream
S	set of lean streams
S_j	the jth lean stream
u_j	fractional saturation of chemically combined A in the jth MSA
x_j	composition of key component in lean stream j, kmol/m³
x_j^s	supply composition of key component in lean stream j, kmol/m³
x_j^t	target composition of key component in lean stream j, kmol/m³
$x_{j,k}$	the upper bound composition for interval k on the scale S_j, kmol/m³
x_j^*	equilibrium composition of key component in lean stream j, kmol/m³
y	composition of key component in any rich stream, kmol/m³
y_i	composition of key component in rich stream i, kmol/m³
y_i^s	supply composition of key component in rich stream i, kmol/m³
y_i^t	target composition of key component in rich stream i, kmol/m³
y^*	equilibrium composition of key component in any rich stream, kmol/m³
z	index for the reactive species

Greek letters

δ_k residual load leaving interval k, kmol/s
$\delta_{i,k}$ residual load leaving interval k for rich stream i, kmol/s
ε_i^R rich-phase minimum allowable composition difference, kmol/m^3
ε_j^S lean-phase minimum allowable composition difference, kmol/m^3
$\nu_{q_j,z,j}$ stoichiometric coefficient of reactive species z in reaction q_j in lean phase j
ξ_{q_j} extent of reaction q_j in the jth MSA

Special symbol

[ ]   concentration, kmol/m^3

References

Astarita, G., and Gioia, F. (1964). Hydrogen sulfide chemical absorption. *Chem. Eng. Sci.*, **19**, 963–971.

Astarita, G., Savage, D. W., and Bisio, A. (1983). "Gas Treating with Chemical Solvents," John Wiley and Sons, New York.

Dunn, R. F., and El-Halwagi, M. M. (1993). Optimal recycle/reuse policies for minimizing the wastes of pulp and paper plants. *J. Environ. Sci. Health*, **A28**(1), 217–234.

El-Halwagi, M. M. (1971). An engineering concept of reaction rate. *Chem. Eng.*, May, 75–78.

El-Halwagi, M. M. (1990). Optimization of bubble column slurry reactors via natural delayed feed addition. *Chem. Eng. Commun.*, **92**, 103–119.

El-Halwagi, M. M., and Srinivas, B. K. (1992), Synthesis of reactive mass-exchange networks. *Chem. Eng. Sci.*, **47**(8), 2113–2119.

Espenson, J. H. (1981). "Chemical Kinetics and Reaction Mechanisms," McGraw-Hill Book Company, New York.

Fogler, S. H. (1992). "Elements of Chemical Reaction Engineering," Prentice Hall, Englewood Cliffs, NJ.

Friedly, J. C. (1991). Extent of reaction in open systems with multiple heterogeneous reactions. *AIChE J.*, **37**(5), 687–693.

Kent, R. L., and Eisenberg, B. (1976). Better data for amine treating. *Hydrocarbon Processing*, February, pp. 87–90.

Kohl, A., and Reisenfeld, F. (1985). "Gas Purification," 4th ed., Gulf Publishing Co., Houston, TX.

Lal, D., Otto, F. D., and Mather, A. E. (1985). The solubility of H$_2$S and CO$_2$ in a diethanolamine solution at low partial pressures. *Can. J. Chem. Eng.*, **63**, 681–685.

Levenspiel, O. (1972). "Chemical Reaction Engineering," 2nd ed., John Wiley and Sons, New York.

Smith, W. R., and Missen, R. W. (1982). "Chemical Reaction Equilibrium Analysis: Theory and Algorithms," John Wiley and Sons, New York.

Srinivas, B. K., and El-Halwagi, M. M. (1994). Synthesis of reactive mass-exchange networks with general nonlinear equilibrium functions. *AIChE J.*, **40**(3), 463–472.

Valenzuela, D. P., and Myers, A. (1989). "Adsorption Equilibrium Data Handbook," Prentice Hall, Englewood Cliffs, NJ.

Warren, A., Srinivas, B. K., and El-Halwagi, M. M. (1995). Design of cost-effective waste-reduction systems for synthetic fuel plants. *J. Environ. Eng.*, **121**(10), 742–747.

Westerterp, K. R., Van Swaaij, W. P. M., and Beenackers, A. A. C. M. (1984). "Chemical Reactor Design and Operation," John Wiley and Sons, New York.

CHAPTER NINE

Combining Heat Integration with Mass Integration

Synthesis of Combined Heat and Reactive Mass-Exchange Networks

In a typical chemical process, there is a strong interaction between mass and energy. For instance, the mass-exchange equilibrium relation of an MSA is affected by its temperature. Therefore, heating or cooling may be beneficial to the performance of the MEN. However, heating/cooling may incur an additional cost. Indeed, there is a trade-off between the mass and energy objectives. Ideally, it is desired to minimize the total cost of heat and mass exchange. In this chapter, we discuss elements of combining heat integration with mass integration. In particular, we present the problem of synthesizing combined heat and reactive mass-exchange networks (CHARMEN) and illustrate how it can be incorporated in a heat-mass integration framework. Before proceeding to the problem of combined heat and mass integration, it is necessary to review the basic principles of synthesizing heat-exchange networks "HENs."

9.1 Synthesis of HENs

In light of the energy crisis in the 1970's, considerable attention has been given to the problem of synthesizing HENs. The problem can be defined as follows:

Given a number N_H of process hot streams (to be cooled) and a number N_C of process cold streams (to be heated), it is desired to synthesize a cost effective network of heat exchangers which can transfer heat from the hot streams to the cold streams. Given also are the heat capacity (flowrate × specific heat) of each process hot stream, $FC_{P,u}$, its supply (inlet) temperature, T_u^s, and target (outlet) temper-

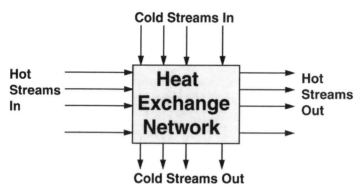

Figure 9.1 Synthesis of HENs.

ature, T_u^t, where $u = 1, 2, \ldots, N_H$. In addition, the heat capacity, $fc_{P,v}$, supply and target temperatures, t_v^s and t_v^t, are given for each process cold stream, where $v = 1, 2, \ldots, N_C$. Available for service are N_{HU} heating utilities and N_{CU} cooling utilities whose supply and target temperatures (but not flowrates) are known. Figure 9.1 illustrates a schematic representation of the HEN problem statement.

For a given system, the synthesis of HENs entails answering several questions including:

• Which heating/cooling utilities should be employed
• What is the optimal heat load to be removed/added by each utility?
• How should the hot and cold streams be matched (i.e., stream pairings)?
• What is the optimal system configuration (e.g., how should the heat exchangers be arranged?, is there any stream splitting and mixing?, etc.)?

Numerous methods have been developed for the synthesis of HENs. These methods have been reviewed by Shenoy (1995), Linnhoff (1993), Gundersen and Naess (1988) and Douglas (1988). One of the key advances in synthesizing HENs is the identification of minimum utility targets ahead of designing the network. The following sections present several methods of determining minimum-utility targets.

9.1.1 Minimum Utility Targets Using the Pinch Diagram

Let us consider a heat exchanger for which the thermal equilibrium relation governing the transfer of the heat from a hot stream to a cold stream is simply given by:

$$T = t. \tag{9.1}$$

By employing a minimum heat-exchange driving force of ΔT^{min}, one can establish a one-to-one correspondence between the temperatures of the hot and the cold streams for which heat transfer is feasible, i.e.,

$$T = t + \Delta T^{min}. \tag{9.2}$$

This expression insures that the heat-transfer considerations of the second law of thermodynamics are satisfied. For a given pair of corresponding temperatures (T, t) it is thermodynamically and practically feasible to transfer heat from any hot stream whose temperature is greater than or equal to T to any cold stream whose temperature is less than or equal to t. It is worth noting the analogy between Eqs. (9.2) and (3.5). Thermal equilibrium is a special case of mass-exchange equilibrium with T, t and $\Delta T^{\min}$ corresponding to y_i, x_j and ε_j, respectively, while the values of m_j and b_j are one and zero, respectively.

In order to accomplish the minimum usage of heating and cooling utilities, it is necessary to maximize the heat exchange among process streams. In this context, one can use a very useful graphical technique referred to as the "thermal-pinch diagram". This technique is primarily based on the work of Linnhoff and co-workers (e.g., Linnhoff and Hindmarsh, 1983), Umeda *et al.* (1979) and Hohmann (1971). The first step in constructing the thermal-pinch diagram is creating a global representation for all the hot streams by plotting the enthalpy exchanged by each process hot stream versus its temperature.[1] Hence, a hot stream losing sensible heat[2] is represented as an arrow whose tail corresponds to its supply temperature and its head corresponds to its target temperature. Assuming constant heat capacity over the operating range, the slope of each arrow is equal to $F_u C_{P,u}$. The vertical distance between the tail and the head of each arrow represents the enthalpy lost by that hot stream according to the following expression:

$$\text{Heat lost from the } u\text{th hot stream “HH}_u\text{”} = F_u C_{p,u}\left(T_u^s - T_u^t\right),$$

where

$$u = 1, 2, \ldots, N_{\text{H}}. \tag{9.3}$$

It is worth noting that any stream can be moved up or down while preserving the same vertical distance between the arrow head and tail and maintaining the same supply and target temperatures. Similar to the graphical superposition described in Chapter Three, one can create a hot composite stream using the diagonal rule. Figure 9.2 illustrates this concept for two hot streams. Next, a cold-temperature scale, t, is created in a one-to-one correspondence with the hot-temperature scale, T, using Eq. (9.2). The enthalpy of each cold stream is plotted versus the cold-temperature scale, t. The vertical distance between the arrow head and tail for a cold stream is given by:

$$\text{Heat gained by the } v\text{th cold stream “HC}_v\text{”} = f_v c_{p,v}\left(t_v^t - t_v^s\right),$$

---

[1] In most HEN literature, the temperature is plotted versus the enthalpy. However, in this chapter enthalpy is plotted versus temperature in order to draw the analogy with MEN synthesis. Furthermore, as will be discussed in Chapter Ten, when there is a strong interaction between mass and energy objectives the enthalpy expressions become nonlinear functions of temperature. In such cases, it is easier to represent enthalpy as a function of temperature.

[2] Whenever there is a change in phase, the latent heat should also be included.

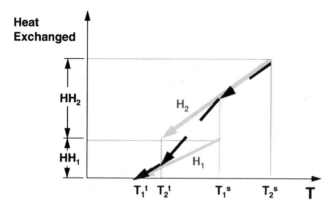

Figure 9.2 Constructing a hot composite stream using superposition (the dashed line represents the composite line).

where

$$v = 1, 2, \ldots, N_C. \tag{9.4}$$

In a similar manner to constructing the hot-composite line, a cold composite stream is plotted (see Fig. 9.3 for a two-cold-stream example).

Next, both composite streams are plotted on the same diagram (Fig. 9.4). On this diagram, thermodynamic feasibility of heat exchange is guaranteed if at any heat-exchange level (which corresponds to a horizontal line), the temperature of the hot composite stream is located to the right of the cold composite stream. Therefore, the cold composite stream can be slid down until it touches the hot composite stream. The point where the two composite streams touch is called

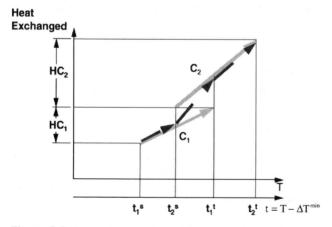

Figure 9.3 Constructing a cold composite stream using superposition (the dashed line represents the composite line).

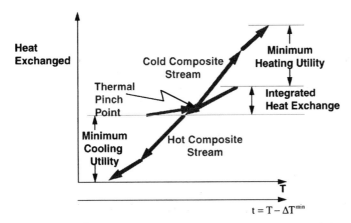

Figure 9.4 The thermal pinch diagram.

the "thermal pinch point." As can be seen from Fig. 9.4, one can use the pinch diagram to determine the minimum heating and cooling utility requirements. The cold composite line cannot be slid down any further, otherwise portions of the cold composite stream become to the right of the hot composite stream, thereby causing thermodynamic infeasibility. On the other hand, if the cold composite stream is moved up (i.e., passing heat through the pinch), less heat integration is possible and, consequently, additional heating and cooling utilities are required. Therefore, for a minimum utility usage no heat should be passed through the pinch. Two additional guidelines are useful for minimizing utility usage; above the pinch only no cooling utilities should be used while below the pinch only no heating utilities should be used. These rules can be explained by noting that above the pinch there is a surplus of cooling capacity. Adding a cooling utility above the pinch will replace a load which can be removed (virtually for no operating cost) by a process cold stream. A similar argument can be made against using a heating utility below the pinch.

9.1.2 Case Study: Pharmaceutical Facility

Consider the pharmaceutical processing facility illustrated in Fig. 9.5. The feed mixture (C_1) is first heated to 550 K then fed to an adiabatic reactor where an endothermic reaction takes place. The off gases leaving the reactor (H_1) at 520 K are cooled to 330 K prior to being forwarded to the recovery unit. The mixture leaving the bottom of the reactor is separated into a vapor fraction and a slurry fraction. The vapor fraction (H_2) exits the separation unit at 380 K and is to be cooled to 300 prior to storage. The slurry fraction is washed with a hot immiscible liquid at 380 K. The wash liquid is purified and recycled to the washing unit. During purification, the temperature drops to 320 K. Therefore, the recycled liquid (C_2) is

Table 9.1

Stream Data for the Pharmaceutical Process

Stream	Flowrate × specific heat kW/°C	Supply temperature, K	Target temperature, K	Enthalpy change kW
H_1	10	520	330	1,900
H_2	5	380	300	400
HU_1	?	560	520	?
C_1	19	300	550	−4,750
C_2	2	320	380	−120
CU_1	?	290	300	?

heated to 380 K. Two utilities are available for service; HU_1 and CU_1. The cost of the heating and cooling utilities ($/10^6$ kJ) are 3 and 5, respectively. Stream data are given in Table 9.1.

In the current operation, the heat exchange duties of H_1, H_2, C_1 and C_2 are fulfilled using the cooling and heating utilities. Therefore, the current annual operating cost of utilities is

$$\left[(4750 + 120)kW \times 3 \times 10^{-6} \frac{\$}{kJ} + (1900 + 400)kW \times 5 \times 10^{-6} \frac{\$}{kJ} \right]$$

$$\times\ 3600 \times 8760 \frac{s}{yr} = \$823,405/yr.$$

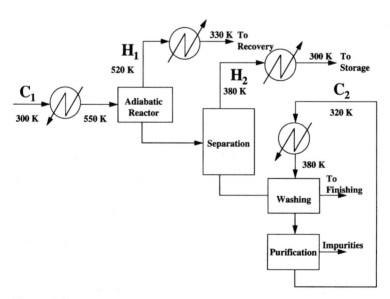

Figure 9.5 Simplified flowsheet for the pharmaceutical process.

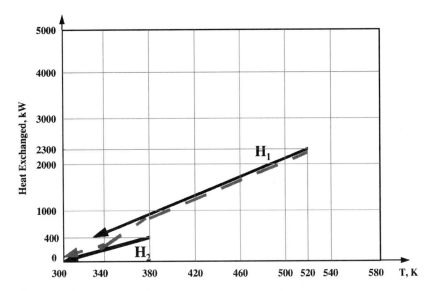

Figure 9.6 The hot composite stream for the pharmaceutical process.

The objective of this case study is to use heat integration via the pinch diagram to reduce this operating cost. A value of $\Delta T^{min} = 10$ K.

Solution Figures 9.6–9.8 illustrate the hot composite stream, the cold composite stream and the pinch diagram, respectively. As can be seen from Fig. 9.8, the two composite streams touch at 310 K on the hot scale (300 K on the cold scale). The minimum heating and cooling utilities are 2,620 and 50 kW, respectively, leading to an annual operating cost of

$$\left(2620\ kW \times 3 \times 10^{-6}\frac{\$}{kJ} + 50\ kW \times 5 \times 10^{-6}\frac{\$}{kJ}\right)3600 \times 8760\frac{s}{yr}$$

$$= \$255,757/yr.$$

This is only 31% of the operating cost prior to heat integration. Once the minimum operating cost is determined, a network of heat exchangers can be synthesized.[3] The trade-off between capital and operating costs can be established by iteratively varying ΔT^{min} until the minimum total annualized cost is attained.

---

[3]Constructing the HEN with minimum number of units and minimum heat-transfer area is beyond the scope of this book. It is analogous to constructing a MEN. The design starts from the pinch following two matching criteria relating number of streams and heat capacities. A detailed discussion on this issue can be found in Linnhoff and Hindmarsh (1983), Douglas (1988), and Shenoy (1995).

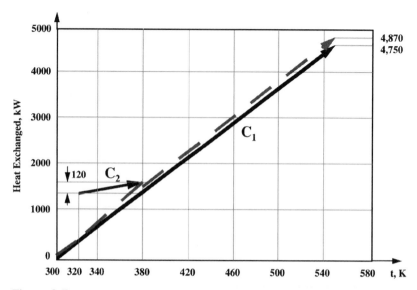

Figure 9.7 The cold composite stream for the pharmaceutical process.

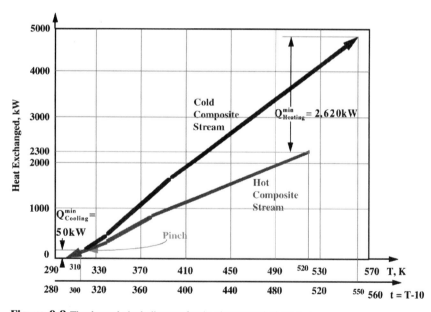

Figure 9.8 The thermal pinch diagram for the pharmaceutical process.

9.1.3 Minimum Utility Targets Using the Algebraic Cascade Diagram

The temperature-interval diagram "TID" is a useful tool for insuring thermodynamic feasibility of heat exchange. It is a special case of the CID described in Chapters Four and Five in which only two corresponding temperature scales are generated; hot and cold. The scale correspondence is determined using Eq. (9.2). Each stream is represented as a vertical arrow whose tail corresponds to its supply temperature while its head represents its target temperature. Next, horizontal lines are drawn at the heads and tails of the arrows. These horizontal lines define a series of temperature intervals $z = 1, 2, \ldots, n_{\text{int}}$. Within any interval, it is thermodynamically feasible to transfer heat from the hot streams to the cold streams. It is also feasible to transfer heat from a hot stream in an interval z to any cold stream which lies in an interval below it.

Next, we construct a table of exchangeable heat loads "TEHL" to determine the heat-exchange loads of the process streams in each temperature interval. The exchangeable load of the uth hot stream (losing sensible heat) which passes through the zth interval is defined as

$$\text{HH}_{u,z} = F_u \text{C}_{\text{p,u}}(\text{T}_{z-1} - \text{T}_z), \tag{9.5}$$

where T_{z-1} and T_z are the hot-scale temperatures at the top and the bottom lines defining the zth interval. On the other hand, the exchangeable capacity of the vth cold stream (gaining sensible heat) which passes through the zth interval is computed through the following expression

$$\text{HC}_{v,z} = f_v c_{p,v}(t_{z-1} - t_z), \tag{9.6}$$

where t_{z-1} and t_z are the cold-scale temperatures at the top and the bottom lines defining the zth interval.

Having determined the individual heating loads and cooling capacities of all process streams for all temperature intervals, one can also obtain the collective loads (capacities) of the hot (cold) process streams. The collective load hot process streams within the zth interval is calculated by summing up the individual loads of the hot process streams that pass through that interval, i.e.,

$$HH_z^{Total} = \sum_{\substack{u \text{ passes through interval } z \\ \text{where } u = 1, 2, \ldots, N_H}} HH_{u,z}. \tag{9.7}$$

Similarly, the collective cooling capacity of the cold process streams within the zth interval is evaluated as follows:

$$HC_z^{Total} = \sum_{\substack{v \text{ passes through interval } z \\ \text{where } v = 1, 2, \ldots, N_C}} HC_{v,z}. \tag{9.8}$$

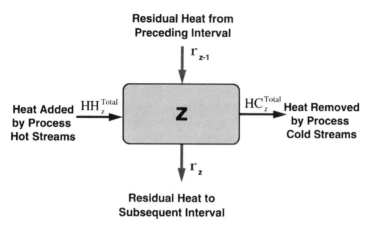

Figure 9.9 A heat balance around a temperature interval.

As has been mentioned earlier, within each temperature interval, it is thermodynamically as well as technically feasible to transfer heat from a hot process stream to a cold process stream. Moreover, it is feasible to pass heat from a hot process stream in an interval to any cold process stream in a lower interval. Hence, for the zth temperature interval, one can write the following heat-balance equation:

$$r_z = HH_z^{Total} - HC_z^{Total} + r_{z-1} \qquad (9.9)$$

where r_{z-1} and r_z are the residual heats entering and leaving the zth interval. Figure 9.9 illustrates the heat balance around the zth temperature interval.

Since no process streams exist above the first interval, r_0 is zero. In addition, thermodynamic feasibility is insured when all the r_z's are nonnegative. Hence, a negative r_z indicates that residual heat is flowing upwards which is thermodynamically infeasible. All negative residual heats can be made non-negative if a hot load equal to the most negative r_z is added to the problem. This load is referred to as the minimum heating utility requirement, $Q_{Heating}^{min}$. Once this hot load is added, the cascade diagram is revised. A zero residual heat designates the thermal-pinch location. The load leaving the last temperature interval is the minimum cooling utility requirement, $Q_{Cooling}^{min}$.

9.1.4 Case Study Revisited Using the Cascade Diagram

We now solve the pharmaceutical case study described in Section 9.1.2. using the algebraic cascade diagram. The first step is the construction of the TID (Fig. 9.10). Next, the TEHL's for the process hot and cold streams are developed (Tables 9.2. and 9.3). Figures 9.11 and 9.12 show the cascade-diagram calculations. The results

Table 9.2
The TEHL for the Process Hot Streams Interval

Interval	Load of H_1, kW	Load of H_2, kW	Total load, kW
1	–	–	–
2	1,300	–	1,300
3	100	–	100
4	500	250	750
5	–	100	100
6	–	50	50

obtained from the revised cascade diagram are identical to those obtained using the graphical pinch approach.

As has been mentioned before, for minimum utility usage no heat should be passed through the pinch. Let us illustrate this point using the cascade diagram. Suppose that we use $Q_{Heating}^{extra}$ kW more than the minimum heating utility. As can be seen from Fig. 9.13, this additional heating utility passes down through the cascade diagram in the form of an increased residual heat load. At the pinch, the residual load becomes $Q_{Heating}^{extra}$. The net effect is not only an increased in the heating utility load but also an equivalent increase in the cooling utility load.

9.1.5 Minimum Utility Targets Using Mathematical Programming (Optimization)

The foregoing algebraic method can be generalized using optimization techniques. A particularly useful approach is the transshipment formulation (Papoulias and

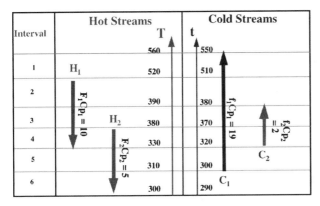

Figure 9.10 The temperature interval diagram for the pharmaceutical case study.

Table 9.3
The TEHL for the Process Cold Streams Interval

Interval	Capacity of C_1, kW	Capacity of C_2, kW	Total capacity kW
1	760	–	760
2	2,470	–	2,470
3	190	20	210
4	950	100	1,050
5	380	–	380
6	–	–	–

Grossmann, 1983). First, we define the loads of heating utilities as follows:

$$HHU_{u,z} = F U_u \, C_{p,u} \, (T_{z-1} - T_z) \quad \text{where}$$

$$u = N_{\mathrm{H}} + 1, N_{\mathrm{H}} + 2, \dots, N_{\mathrm{H}} + N_{\mathrm{HU}} \quad (9.10)$$

and

$$HHU_z^{Total} = \sum_{\substack{u \text{ passes through interval } z \\ \text{where } u = N_{\mathrm{H}} +1, N_{\mathrm{H}} +2, \dots, N_{\mathrm{H}} + N_{\mathrm{HU}}}} HUU_{u,z}. \quad (9.11)$$

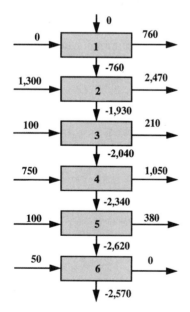

Figure 9.11 The cascade diagram for the pharmaceutical case study.

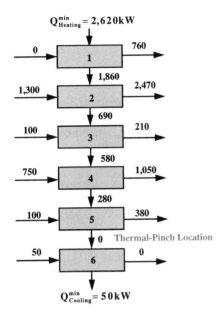

$$Q^{min}_{Heating} = 2,620 kW$$

0 → [1] → 760
1,860
1,300 → [2] → 2,470
690
100 → [3] → 210
580
750 → [4] → 1,050
280
100 → [5] → 380
0   Thermal-Pinch Location
50 → [6] → 0
$$Q^{min}_{Cooling} = 50 kW$$

Figure 9.12 The revised cascade diagram for the pharmaceutical case study.

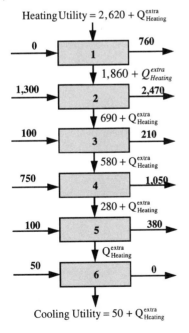

$$Heating\, Utility = 2,620 + Q^{extra}_{Heating}$$

0 → [1] → 760
$$1,860 + Q^{extra}_{Heating}$$
1,300 → [2] → 2,470
$$690 + Q^{extra}_{Heating}$$
100 → [3] → 210
$$580 + Q^{extra}_{Heating}$$
750 → [4] → 1,050
$$280 + Q^{extra}_{Heating}$$
100 → [5] → 380
$$Q^{extra}_{Heating}$$
50 → [6] → 0
$$Cooling\, Utility = 50 + Q^{extra}_{Heating}$$

Figure 9.13 Consequences of passing heat through the pinch.

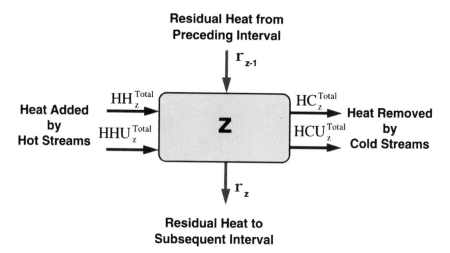

Figure 9.14 A heat balance around a temperature interval including utilities.

Similarly, the cooling capacities of the cold utilities through the following expression:

$$HCU_{v,z} = fU_v c_{p,v}(t_{z-1} - t_z) \quad \text{where}$$

$$v = N_C + 1, N_C + 2, \ldots, N_C + N_{CU} \quad (9.12)$$

and

$$HCU^{Total} = \sum_{\substack{v \ passes \ through \ interval \ z \\ where \ v = N_C +1, N_C +2, \ldots, N_C + N_{CU}}} HCU_{v,z}. \quad (9.13)$$

For the zth temperature interval, one can write the following heat-balance equation (see Fig. 9.14):

$$HH_z^{Total} - HC_z^{Total} = HHU_z^{Total} - HHC_z^{Total} + r_{z-1} - r_z,$$

$$z = 1, 2, \ldots n_{int}. \quad (9.14)$$

where

$$r_0 = r_{n_{int}} = 0 \quad (9.15)$$

$$r_z \geq 0, \quad z = 1, 2, \ldots, n_{int} - 1 \quad (9.16)$$

$$FU_u \geq 0, \quad u = N_H + 1, N_H + 2, \ldots, N_H + N_{HU} \quad (9.17)$$

$$fU_v \geq 0 \quad v = N_C + 1, N_C + 2, \ldots, N_C + N_{CU}. \quad (9.18)$$

Now, the objective function of utility cost can be minimized subject to the set of constraints (9.10)–(9.18). This formulation is a linear program which can be solved using commercially available software (e.g., LINGO).

9.1.6 Case Study Revisited Using Linear Programming

We are now in a position to solve the pharmaceutical case study (Section 9.1.2) using optimization techniques. The first step is to create the TID including process streams and utilities (Fig. 9.15). Next, the problem is formulated as an optimization program as follows:

$$\text{minimize}\left(3 \times 10^{-6} Q_{Heating}^{min} + 5 \times 10^{-6} Q_{Cooling}^{min}\right)3,600 \times 8,760$$

$$= 94.608 Q_{Heating}^{min} + 157.68 Q_{Cooling}^{min}$$

Subject to

$$r_1 - Q_{Heating}^{min} = -760$$

$$r_2 - r_1 = -1,170$$

$$r_3 - r_2 = -110$$

$$r_4 - r_3 = -300$$

$$r_5 - r_4 = -280$$

$$-r_5 + Q_{Cooling}^{min} = 50$$

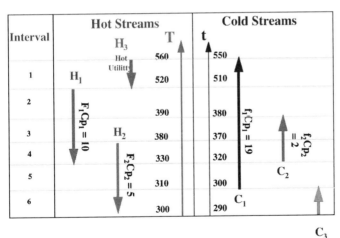

Figure 9.15 The temperature interval diagram for the pharmaceutical case study.

with all the non-negativity constraints. In terms of LINGO input, the program can be written as follows:

```
model:
min = 94.608*QHmin + 157.68*QCmin;
r1 - QHmin = -760;
r2 - r1 = -1170;
r3 - r2 = -110;
r4 - r3 = -300;
r5 - r4 = -280;
-r5 + QCmin = 50;
QHmin >= 0;
QCmin >= 0;
r1 >= 0;
r2 >= 0;
r3 >= 0;
r4 >= 0;
r5 >= 0;
```

The solution obtained using LINGO is:

```
Objective value:       255757.0
       Variable           Value
        QHMIN           2620.000
        QCMIN           50.00000
          R1            1860.000
          R2            690.0000
          R3            580.0000
          R4            280.0000
          R5            0.0000000E+00
```

These results are identical to those obtained using the graphical and the algebraic methods.

9.2 Synthesis of Combined Heat- and Reactive Mass-Exchange Networks "CHARMEN"

Heretofore, the presented MEN/REAMEN synthesis techniques were applicable to the cases where mass-exchange temperatures are known ahead of the synthesis task. Mass-exchange equilibrium relations are dependent upon temperature and the selection of optimal mass-exchange temperatures is an important element of design. In selecting these temperatures, there is a tradeoff between cost of MSAs and cost of heating/cooling utilities.

The CHARMEN synthesis problem can be stated as follows: Given a number N_R of waste (rich) streams and a number N_S of lean streams (physical and reactive MSAs), it is desired to synthesize a cost-effective network of physical and/or reactive mass exchangers which can preferentially transfer certain undesirable species from the waste streams to the MSAs. Given also are the flowrate of each waste stream, G_i, its supply (inlet) composition, y_i^s, and target (outlet) composition, y_i^t, and the supply and target compositions, x_j^s and x_j^t for each MSA. In addition, available for service are hot and cold streams (process streams as well as utilities) that can be used to optimize the mass-exchange temperatures.

Due to the interaction of mass and heat in the CHARMEN, one should synthesize the MEN and the HEN simultaneously. This section presents an optimization based method for the synthesis of CHARMEN's[4]. Two key assumptions are invoked:

• Each mass exchanger operates isothermally. While each lean or rich stream can assume several values of temperature that differ from one mass exchanger to another, it is assumed that within each mass exchanger each stream passes isothermally.

• In the range of operating temperatures and compositions, the equilibrium relations are monotonic functions of temperature of the MSA. This is typically true. For instance, normally in gas absorption Henry's coefficient monotonically decreases as the temperature of the MSA is lowered while for stripping the gas-liquid distribution coefficient monotonically increases as the temperature of the stripping agent is increased.

The concept of "lean substreams" (El-Halwagi, 1992/1993; Srinvias and El-Halwagi, 1994) is useful in trading off mass versus heat objectives. Each MSA, j, is assumed to split into ND_j lean substreams. A lean substream S_{j,d_j} is defined as the d_jth (where $d_j = 1, 2, \ldots, ND_j$) portion of MSA j whose composition and temperature vary between a supply value of (x_j^s, T_j^s) and a target value of (x_j^t, T_j^t) and has a flowrate of L_{j,d_j} which does not split or mix with other substreams. It can be shown that a MOC solution of a CHARMEN is realized when each lean substream exchanges mass at a single temperature T_{j,d_j}^* (for the proof, the reader is referred to Srinvias and El-Halwagi, 1994). The identification of temperature T_{j,d_j}^* and flowrate L_{j,d_j} for each lean substream is part of the optimization problem.

In setting up the CHARMEN-synthesis formulation, we first choose a number ND_j of lean substreams for MSA j, each operating at a selected temperature T_{j,d_j}^* which lies within the admissible temperature range for the MSA. The number of substreams is dependent on the level of accuracy needed for equilibrium dependence on temperature. Theoretically, an infinite number of substreams should

---

[4]More rigorous techniques for optimizing outlet composition are described by Srinvias and El-Halwagi (1994).

be used to cover the whole temperature span of each MSA. However, in practice few (typically less than five) substreams are needed. Since each temperature corresponds to an equilibrium relation, each lean substream should have its own composition scale on the CID. On the CID, the various substreams are represented against their composition scale. For each lean substream, two heat exchange tasks take place; one before mass exchange and one after. First, the temperature of the leans substream is changed from T_j^s to T_{j,d_j}^*. Mass exchange occurs at T_{j,d_j}^*. Then, the temperature of the substream has to be altered from T_{j,d_j}^* to T_j^i. These are heat-exchange tasks that can be handled as discussed in Section 9.1.

Therefore, the CHARMEN synthesis problem can be formulated by combining the HEN formulation of Section 9.1 and the REAMEN synthesis equations developed in Chapter Eight after adjustment to incorporate the notion of substreams. For instance, the cost of MSAs can be expressed as follows:

$$Cost\ of\ MSAs = \sum_{j=1}^{N_s} C_j \sum_{d_j=1}^{ND_j} L_{j,d_j} \qquad (9.19)$$

and the material balances around the composition intervals become:

$$\delta_k - \delta_{k-1} + \sum_{j} \sum_{d_j\ \text{passes through interval k}} L_{j,d_j} w_{j,k}^S = W_k^R, \quad k = 1, 2, \ldots, N_{int},$$
$$(9.20)$$

where

$$L_{j,d_j} \geq 0, \quad j = 1, 2, \ldots, N_S \qquad (9.21)$$

and

$$\sum_{d_j=1}^{ND_j} L_{j,dj} \leq L_j^C, \quad j = 1, 2, \ldots, N_S \qquad (9.22)$$

with the mass-residual constraints:

$$\delta_0 = \delta_{N_{int}} = 0 \qquad (9.23)$$
$$\delta_k \geq 0 \quad k = 1, 2, \ldots, N_{int} - 1 \qquad (9.24)$$

The above equations coupled with Eqs. (9.10)–(9.18) represent the constraints of the CHARMEN-synthesis formulation. The objective is to minimize the cost of MSAs and heating/cooling utilities. This is a linear-programming formulation whose solution determines the optimal flowrate and temperature of each substream and heating/cooling utilities. In order to demonstrate this formulation, let us consider the following example.

Table 9.4
Data of Waste Stream for the Ammonia Removal Example

Stream	Supply flowrate G_i, kg/s	Supply composition (mass fraction) y_i^s	Target Composition (mass fraction) y_i^t
R_1	1.0	0.011	0.001

9.3 Case Study: CHARMEN Synthesis for Ammonia Removal from a Gaseous Emission

A gaseous emission is to be treated for the removal of ammonia. Table 9.4 provides the stream data. Two scrubbing agents are considered for the removal of ammonia, water, S_1, and an inorganic solvent, S_2. The absorption of ammonia in water is coupled with the following chemical reaction:

$$NH_3 + H_2O = NH_4^+ + OH^- \qquad (9.25)$$

Within the considered range of operation, the equilibrium relation for ammonia scrubbing in water is dependent on the temperature as follows:

$$y = (0.053T_1 - 14.5)x_1 \qquad (9.26)$$

where y is the mass fraction of ammonia in the gas, T_1 is the water temperature in K and x_1 is the mass fraction of ammonia in the water. The data for the lean streams are summarized in Table 9.5.

Table 9.5
Data of the Lean Streams for the Ammonia Removal Example

Stream	Supply composition of ammonia (mass fraction) x_j^s	Target composition of ammonia (mass fraction) x_j^t	m_j	ϵ_j	C_j	T_j^s (K)	T_j^t (K)	T_j^{LB} (K)	T_j^{UB} (K)
S_1	0.0000	0.0050	0.053T-14.5 (where T is in K)	0.0010	0.005	298	$\leq$298	283	300
S_2	0.0040	0.1090	0.100	0.0010	0.160	295	295	295	295

Table 9.6
Stream Data for the Coolants

Stream	Supply temperature, K	Target temperature, K	Cost $/kg coolant
C_1	288	293	0.002
C_2	278	283	0.006

Two cooling utilities are available; CU_1 and CU_2. The specific heat for both coolants is assumed to be that of water. A value of $\Delta T^{\min} = 5\ K$ is used. The data for the two cold streams are given in Table 9.6.

The objective of this case study is to synthesize a CHARMEN which has a minimum operating cost (cost of MSAs + cost of cooling utilities).

In order to cover the temperature span for water, let us choose four substreams operating at 283, 288, 293 and 298 K. At each temperature, the equilibrium constant is evaluated. The CID for the problem is shown by Fig. 9.16.

The objective function is to minimize the operating cost of the MSAs and the cooling utilities, i.e.,

$$\min[0.005(L_{1,1} + L_{1,2} + L_{1,3} + L_{1,4}) + 0.160L_2 + 0.002fU_1 + 0.006fU_2]$$

$$\times 3600 \times 8760$$

$$= 157{,}680(L_{1,1} + L_{1,2} + L_{1,3} + L_{1,4}) + 5{,}045{,}760L_2 + 63{,}072fU_1$$

$$+189{,}216fU_2,$$

subject to the following constraints.

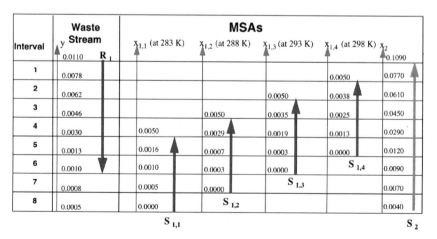

Figure 9.16 The composition interval diagram for ammonia removal case study.

Material balances around composition intervals

$$\delta_1 + 0.0320L_2 = 0.0032$$

$$\delta_2 - \delta_1 + 0.0012L_{1,4} + 0.0160L_2 = 0.0016$$

$$\delta_3 - \delta_2 + 0.0015L_{1,3} + 0.0013L_{1,4} + 0.0160L_2 = 0.0016$$

$$\delta_4 - \delta_3 + 0.0021L_{1,2} + 0.0016L_{1,3} + 0.0012L_{1,4} + 0.0160L_2 = 0.0016$$

$$\delta_5 - \delta_4 + 0.0034L_{1,1} + 0.0022L_{1,2} + 0.0016L_{1,3} + 0.0013L_{1,4}$$
$$+ 0.0170L_2 = 0.0017$$

$$\delta_6 - \delta_5 + 0.0005L_{1,1} + 0.0004L_{1,2} + 0.0003L_{1,3} + 0.0030L_2 = 0.0003$$

$$\delta_7 - \delta_6 + 0.0005L_{1,1} + 0.0003L_{1,2} + 0.0020L_2 = 0.0000$$

$$-\delta_7 + 0.0005L_{1,1} + 0.0030L_2 = 0.0000.$$

Heat balances around composition intervals

$$r_1 + 5.0 \times 4.18fU_1 - 5.0 \times 4.18(L_{1,1} + L_{1,2} + L_{1,3}) = 0.0$$

$$-r_2 + 5.0 \times 4.18fU_2 - 5.0 \times 4.18L_{1,1} = 0.0$$

$$r_2 - r_1 - 5.0 \times 4.18(L_{1,1} + L_{1,2}) = 0.0.$$

These equations coupled with the non-negativity constraints form a linear program which can be modeled on LINGO as follows:

```
MODEL:
MIN = 157680*L11 + 157680*L12 + 157680*L13
 + 157680*L14 + 5045760*L2 + 63072*FU1 + 189216*FU2;

D1 + 0.0320*L2 = 0.00320;
D2 - D1 + 0.0012*L14+0.0160*L2 = 0.00160;
D3 - D2 + 0.0015*L13+0.0013*L14+0.0160*L2=0.00160;
D4 - D3 + 0.0021*L12 + 0.0016*L13 + 0.0012*L14
 + 0.0160*L2 = 0.00160;
D5 - D4 + 0.0034*L11 + 0.0022*L12 + 0.0016*L13
 + 0.0013*L14 + 0.0170*L2 = 0.00170;
D6 - D5 + 0.0005*L11 + 0.0004*L12 + 0.0003*L13
 + 0.0030*L2 = 0.00030;
D7-D6+0.0005*L11+0.0003*L12+0.0020*L2=0.00000;
 - D7 + 0.0005*L11 + 0.0030*L2 = 0.00000;

r1 + 20.90*FU1 - 20.90*(L11 + L12 + L13) = 0.0;
r2 - r1 - 20.90*(L11 + L12) = 0.0;
```

```
- r2 + 20.90*FU2 - 20.90*L11 = 0.0;
D1 >= 0.0;
D2 >= 0.0;
D3 >= 0.0;
D4 >= 0.0;
D5 >= 0.0;
D6 >= 0.0;
D7 >= 0.0;
D8 >= 0.0;
D9 >= 0.0;
r1 > 0.0;
r2 > 0.0;
r3 > 0.0;
L11 >= 0.0;
L12 >= 0.0;
L13 >= 0.0;
L14 >= 0.0;
L2 >= 0.0;
FU1 >= 0.0;
FU2 >= 0.0;
end
```

```
Objective value:          378432.0
            Variable         Value
              L11       0.0000000E+00
              L12       0.0000000E+00
              L13       1.000000
              L14       0.9999999
              L2        0.0000000E+00
              FU1       1.000000
              FU2       0.0000000E+00
              D1        0.3200000E-02
              D2        0.3600000E-02
              D3        0.2400000E-02
              D4        0.1200000E-02
              D5        0.0000000E+00
              D6        0.0000000E+00
              D7        0.0000000E+00
              R1        0.0000000E+00
              R2        0.0000000E+00
              D8        0.0000000E+00
              D9        0.0000000E+00
              R3        0.0000000E+00
```

Interval	Hot Streams $S_{1,1}$ $S_{1,2}$ $S_{1,3}$	T (K)	t (K)	Cold Streams
		298	293	
1		293	288	
2		288	283	CU_1
3		283	278	
				CU_2

Figure 9.17 The temperature interval diagram for the ammonia removal case study.

The solution involves the use of two water substreams for scrubbing; at 293 and 298 K. The first coolant is employed to undertake the cooling duty. Several conceptual flowsheets can be developed to implement this solution. Figures 9.18a and b illustrate such configurations. The developed mass exchanger is composed two sections. The top section employs $S_{1,3}$ to reduce the ammonia composition in the gas from 0.13% to 0.10% and operates at an average water temperature of 293 K. In Fig. 9.18a, temperature is reduced by indirect contact cooling of the

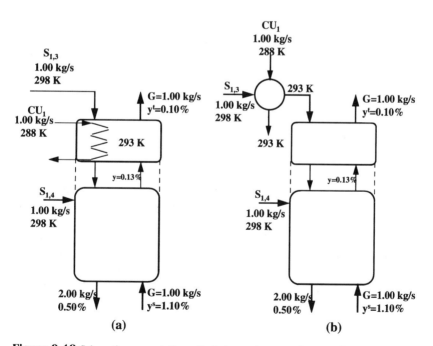

Figure 9.18 Schematic representations of solution to the ammonia removal case study: (a) with internal cooling, (b) with external cooling.

column internals. On the other hand, Fig. 9.18b uses a heat exchanger to cool $S_{1,3}$ to 293 K. The rest of the column achieves the remaining separation (1.10 to 0.13% ammonia in gas) and involves the use of $S_{1,4}$ and $S_{1,3}$. It is worth pointing out that following the synthesis task, an analysis study is needed to size the column, predict the actual temperature profile in the column and validate or refine the water flow and system performance.

9.4 Case Study: Combining Heat and Mass Integration for an Ammonium Nitrate Plant

Ammonium nitrate is manufactured by reacting ammonia with nitric acid. Consider the process shown by Fig. 9.19. First, natural gas is reformed and converted into hydrogen, nitrogen and carbon dioxide. Hydrogen and nitrogen are separated an fed to the ammonia synthesis plant. A fraction of the produced ammonia is employed in nitric acid formation. Ammonia is first oxidized with compressed air then absorbed in water to form nitric acid. Finally nitric acid is reacted with ammonia to produce ammonium nitrate.

The plant disposes of two waste streams; gaseous and aqueous. The gaseous emission results from the ammonia and the ammonium nitrate plants. It is fed to an incinerator prior to atmospheric disposal. In the incinerator, ammonia is converted into NO_x. Due to more stringent NO_x regulations, the composition of ammonia in the feed to the incinerator has to be reduced from 0.57 wt% to 0.07 wt%. The lean streams presented in Table 9.5 may be employed to remove ammonia. The main aqueous waste of the process results from the nitric acid plant. Due to its acidic content of nitric acid, it is neutralized with an aqueous ammonia solution before biotreatment.

The objective of this case study is identify a cost-effective solution that can reduce the ammonia content of the feed to the incinerator to 0.07 wt%.

Solution

Let us first segregate the two sources forming the feed to the incinerator. As can be seen from the source-sink mapping diagram (Fig. 9.20), the gaseous emission from the ammonium nitrate process (R_2) is within the acceptable zone for the incinerator. Therefore, it should not be mixed with R_1 then separated. Instead, the ammonia content of R_1 should be reduced to 0.10 wt% then mixed with R_2 to provide an acceptable feed to the incinerator as shown by Fig. 9.20. The task of removing ammonia from R_1 to from 1.10 wt% to 0.10 wt% is identical to the case study solved in Section 9.3. Hence, the solution presented in Fig. 9.18 can be used.

Instead of disposing the water leaving the scrubber for R_1 into the biotreatment facility, it is beneficial to use it in a process sink. An appropriate sink is the mixer preceding the neutralizer (used in neutralizing the wastewater from

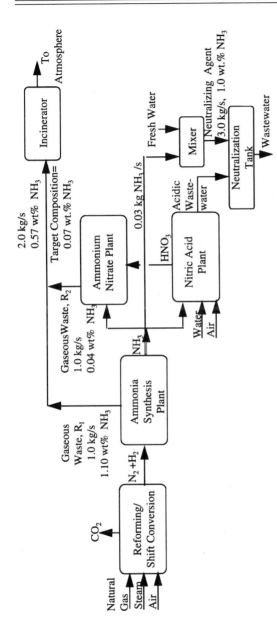

Figure 9.19 A simplified flowsheet for the manufacture of ammonium nitrate.

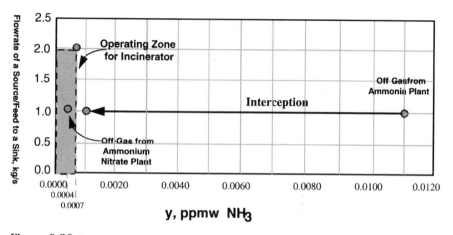

Figure 9.20 A gaseous source-sink mapping diagram for the ammonium nitrate example.

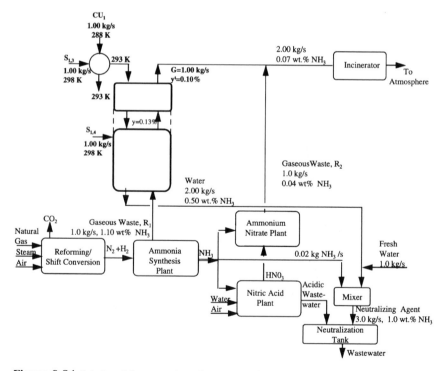

Figure 9.21 Solution of the ammonium nitrate case study.

Table 9.7
Stream Data for Problem 9.1 (Douglas, 1988)

Stream	Flowrate × specific heat Btu/hr °F	Supply temperature, °F	Target temperature, °F	Enthalpy change 10^3 Btu/hr
H_1	1,000	250	120	130
H_2	4,000	200	100	400
HU_1	?	280	250	?
C_1	3,000	90	150	-180
C_2	6,000	130	190	-360
CU_1	?	60	80	?

the nitric acid). Hence, the water leaving the R_1 scrubber can be used to replace the fresh-water consumption in the mixer by 2.0 kg/s (173 tons/day). Due to its ammonia content (0.50 wt.%), the amount of ammonia needed for neutralization is reduced by 0.01 kg/s. For an ammonia cost of $100/ton, the ammonia savings are about $32,000/year. Figure 9.21 illustrates the identified solution.

Problems

9.1 A plant has two process hot streams (H_1 and H_2), two process cold streams (C_1 and C_2), a heating utility (HU_1) and a cooling utility (CU_1). The problem data are given in Table 9.7. Using graphical, algebraic and optimization techniques determine the minimum heating and cooling requirements for the problem. The minimum allowable temperature difference is 10°F.

9.2 Consider a process which has two process hot streams (H_1 and H_2), two process cold streams (C_1 and C_2), a heating utility (HU_1 which is a saturated vapor that loses its latent heat of condensation) and a cooling utility (CU_1). The problem data are given in Table 9.8. The cost of the heating utility is $4/10^6$ kJ added while the cost of the coolant is $7/10^6$ kJ. The minimum allowable temperature difference is 10°C. Employ graphical, algebraic and

Table 9.8
Stream Data for Problem 9.2 (Papoulias and Grossmann, 1983)

Stream	Flowrate × specific heat kW/ °C	Supply temperature, °C	Target temperature, °C
H_1	10.55	249	138
H_2	8.79	160	93
HU_1	?	270	270
C_1	7.62	60	160
C_2	6.08	116	260
CU_1	?	38	82

Table 9.9
*Supply and Target Compositions for the MSAs
in the Pulping Case Study*

Stream	Flowrate kg/s	y_i^s (mass fraction)	y_i^t (mass fraction)
R_1	104	8.83×10^{-4}	5.00×10^{-6}
R_2	442	7.00×10^{-4}	5.00×10^{-6}

optimization techniques determine the minimum heating and cooling requirements for the process.

9.3 Consider a pulping process which results in two gaseous emissions that are rich in H_2S (R_1 and R_2) that requires to be desulfurized. The task of desulfurizing the two gaseous streams within the plant may be accomplished by the use of one or more of the following reactive mass-exchange processes: reactive absorption of H_2S in the white liquor (S_1) to produce Na_2S (a necessary ingredient for kraft pulping), absorption in N-mehtl-2-pyrrolidone (S_2) and absorption in methyldiethanolamine "MDEA" (S_3). The operating cost of MSAs (primarily regeneration and makeup) is \$0.000, 0.007 and 0.001/kg for S_1, S_2 and S_3, respectively. Also, heating/cooling utilities are available in the form of heating oil, cooling water and chilled nitrogen with costs of 3.63×10^{-3}, 1.3×10^{-5}, and 3.10×10^{-3} \$/kg, respectively. The equilibrium data for H_2S in the MSAs as a function of temperature can be found in literature (Srinivas and El-Halwagi, 1994). The minimum allowable composition differences, ε_j, for MSAs S_1, S_2 and S_3 are 1.0×10^{-3} 1.0×10^{-6}, and 1.0×10^{-3}, respectively. Thermodynamic feasibility for heat exchange was ensured by minimum allowable temperature difference, ΔT^{min}, of 10 K. Stream data are given in Tables 9.9–9.11. Synthesize an optimum CHARMEN that can transfer the H_2S from the waste streams (R_1 and R_2) to one or more of the MSAs (S_1–S_3).

Symbols

C_j unit cost of the jth MSA including regeneration and makeup, \$/kg of recirculating MSA

Table 9.10
Data for the Waste Streams in the Pulping Case Study

MSA	Description	L_j^c kg/s	x_j^s (mass fraction)	x_j^t (mass fraction)
S_1	White liquor	40	0.07557	0.115
S_2	15 wt% N-methyl-2-pyrrolidone	∞	1.0×10^{-6}	1.0×10^{-5}
S_3	15 wt% MDEA	∞	0.001	0.010

Table 9.11
Thermal Data for the Streams in the Pulping Case Study

Stream	Supply temperature K	Target temperature K	Lower-bound temperature K	Upper-bound temperature K	Average specific heat kJ/kg K
R_1	298	298	288	313	1.00
R_2	298	298	288	313	1.00
S_1	368	368	368	368	2.50
S_2	313	313	298	313	2.40
S_3	310	310	280	330	2.40
Heating oil	373	343	343	373	2.2
Cooling water	288	298	288	298	4.2
Chilled nitrogen	100	100	100	100	

$C_{P,u}$	Specif heat of hot stream u, kJ/(kg K)
$c_{P,v}$	Specif heat of cold stream v, kJ/(kg K)
dj	index for substreams of the jth MSA
f	flowrate of cold stream, kg/s
fU_v	flowrate of cold utility v, kg /s
F	flowrate of hot stream, kg/s
FU_u	flowrate of hot utility u, kg /s
G_i	flowrate of the ith waste stream
$HCU_{v,z}$	cold load in interval z, defined by Eq. (9.12)
HCU_z^{Total}	total cold load in interval z, defined by Eq. (9.13)
$HHU_{u,z}$	hot load in interval z, defined by Eq. (9.10)
HHU_z^{Total}	total hot load in interval z, defined by Eq. (9.11)
i	index of waste streams
j	index of MSAs
k	index of composition intervals
L_j	flowrate of the jth MSA, kg/s
L_{j,d_j}	flowrate of substream of dj the jth MSA, kg/s
L_j^c	upper bound on available flowrate of the jth MSA, kg/s
m_j	slope of equilibrium line for the jth MSA
N_C	number of process cold streams
N_{CU}	number of cooling utilities
N_H	number of process hot streams
N_{HU}	number of process cold streams
N_{int}	number of composition intervals
N_R	number of waste streams
N_S	number of MSAs
ND_j	Number of substreams for the jth MSA
R_i	the ith waste stream
S_j	the jth MSA

$S_{j,dj}$ the d_jth substream of the jth MSA
t temperature of cold stream, K
t_v^s supply temperature of cold stream v, K
t_v^t target temperature of cold stream v, K
T_u^s supply temperature of hot stream u, K
T_u^t target temperature of hot stream u, K
T temperature of hot stream, K
u index for hot streams
v index for cold streams
$W_{i,k}^R$ exchangeable load of the ith waste stream which passes through the kth interval, kg/s
$w_{j,k}^S$ exchangeable load of the jth MSA which passes through the kth interval, kg/s
W_k^R the collective exchangeable load of the waste streams in interval k, kg/s
$x_{j,k}$ composition of key component in the jth MSA at the lower horizontal line defining the kth interval
x_j^s supply composition of the jth MSA
x_j^t target composition of the jth MSA
z temperature interval

Greek
δ_k mass residual from interval k, kg/s
$\Delta T^{\min}$ minimum approach temperature, K
ε_j minimum allowable composition difference for the jth MSA

References

Douglas, J. M. (1988). "Conceptual Design of Chemical Processes," pp. 216–288, McGraw Hill, New York.

El-Halwagi, M. M. (1992). A process synthesis approach to the dilemma of simultaneous heat recovery, waste reduction and cost effectiveness. In "Proceedings of the Third Cairo International Conference on Renewable Energy Sources," (A. I. El-Sharkawy and R. H. Kummler, eds.) Vol. 2, pp. 579–594.

Gundersen, T., and Naess, L. (1988). The synthesis of cost optimal heat exchanger networks: An industrial review of the state of the art. Comp. Chem. Eng., 12(6), 503–530.

Hohmann, E. C. (1971). Optimum networks for heat exchanger. Ph.D. Thesis, University of Southern California, Los Angeles.

Linnhoff, B. (1993). Pinch analysis—A state of the art overview. Trans. Inst. Chem. Eng. Chem. Eng. Res. Des. 71, Part A5, 503–522.

Linnhoff, B., and Hindmarsh, E. (1983). The pinch design method for heat exchanger networks, Chem. Eng. Sci. 38(5), 745–763.

Papoulias, S. A., and Grossmann, I. E. (1983). A structural optimization approach in process synthesis-II. Heat recovery networks, Comput. Chem. Eng. 7(6), 707–721.

Shenoy, U. V. (1995). "Heat Exchange Network Synthesis: Process Optimization by Energy and Resource Analysis." Gulf Publ. Co., Houston, TX.

Srinivas, B. K., and El-Halwagi, M. M. (1994). Synthesis of combined heat reactive mass-exchange networks, *Chem. Eng. Sci.* **49**(13), 2059–2074.

Umeda, T., Itoh, J., and Shiroko, K. (1979). A thermodynamic approach to the synthesis of heat integration systems in chemical processes. *Comp. Chem. Eng.* **3**, 273–282.

CHAPTER TEN

Synthesis of Heat-Induced Separation Network for Condensation of Volatile Organic Compounds

The previous chapters have focused on separation technologies that employ mass-separating agents. Another important class of separations involves the use of energy-separating agents "ESAs" to induce separations. This chapter deals the optimal design of *heat-induced separation networks* "HISENs". Examples of HISENs include condensation, crystallization and drying. The chapter will focus on condensation due to its importance in recovering volatile organic compounds "VOCs" which are among the most serious atmospheric pollutants. A shortcut graphical method will be presented. This method is primarily based on the work of Richburg and El-Halwagi (1995). More generalized problem statements and design procedures can be found in literature (e.g. Dunn and El-Halwagi, 1994a,b; El-Halwagi *et al.*, 1995; Dunn *et al.*, 1995).

10.1 Problem Statement

Given a VOC-laden gaseous stream whose flowrate is G, supply composition of the VOC is y^s and supply temperature is T^s, it is desired to design a cost-effective condensation system which can recover a certain fraction, α, of the amount of the VOC contained in the stream.

Available for service are several refrigerants. The operating temperature for the jth refrigerant, t_j, is given. For convenience in terminology, these refrigerants are arranged in order of decreasing operating temperature, i.e.,

$$t_1 \geq t_2 \geq \ldots t_j \ldots \geq t_{NE} \qquad (10.1)$$

The operating cost of the jth refrigerant (denoted by C_j, \$/kJ removed) is known. The flowrate of each refrigerant is unknown and is to be determined through optimization.

10.2 System Configuration

Normally, the VOC-condensation system involves three units as schematically depicted in Fig. 10.1. The initial step is to cool the stream to a temperature slightly above that of water freezing so as to dehumidify the gas by condensing the water vapor and prevent the detrimental icing effects in subsequent stages. Next, the stream is cooled to T^* to recover the VOC. The temperature T^* is an optimization variable. In order to utilize the cooling capacity of the gaseous stream at T^*, it is recycled back to the system for heat integration. The remaining cooling duty is accomplished by a refrigerant.

It is worth pointing out that the bypassed portion in Fig. 10.2 is an optimization variable. The bypass is strongly linked to the selection of a target composition for the VOC in the outlet gas. At first glance, it may appear that the target composition is calculated via

$$y^t = (1 - \alpha)y^s, \tag{10.2}$$

which is not necessarily optimal. Indeed, one may pass only a fraction, β, of the gaseous stream through the condensation system such that the amount condensed from the fraction β is equal to the amount to be recovered from the whole stream, i.e.,

$$\beta G(y^s - y^t) = \alpha G y^s \tag{10.3a}$$

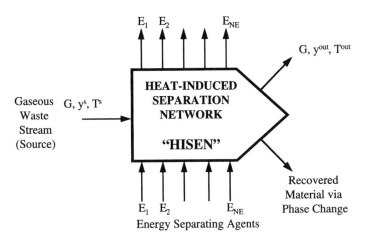

Figure 10.1 Schematic representation of the HISEN synthesis problem.

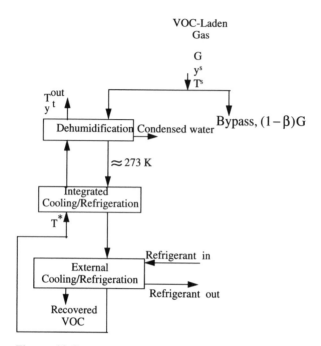

Figure 10.2 A VOC-recovery system (from Richburg and El-Halwagi, 1995. Reproduced with permission of the American Institute of Chemical Engineers. Copyright © 1995 AIChE. All rights reserved).

or

$$y^t = \left(1 - \frac{\alpha}{\beta}\right)y^s. \tag{10.3b}$$

since condensation must occur ($y^t < y^s$) the following bounds on β should be met:

$$\alpha < \beta \leq 1.00 \tag{10.4}$$

The rest of the gaseous stream, $(1 - \beta)G$, is bypassed, and the net effect is that a fraction α of the VOC contained in the whole gaseous waste is recovered. Hence, the identification of the optimum value of β is part of the system optimization.

10.3 Integration of Mass and Heat Objectives

The primary objective of the VOC-condensation system is to meet mass-recovery objectives. However, heat is a key element in realizing the mass objectives. Hence, the mass and heat interactions of the problem have to be identified and reconciled. This can be achieved by converting the VOC-recovery task from a mass-transfer problem to a heat-transfer duty. This can be accomplished by relating the composition of the VOC to the temperature of the gaseous waste. When a VOC-laden gas is cooled, the composition of the VOC remains constant until condensation starts at a temperature T^c, defined by

$$p^s = p^o(T^c), \qquad (10.5)$$

where p^s is the supply (inlet) partial pressure of the VOC, $p^o(T)$ is the vapor pressure of the VOC expressed as a function of the gaseous-waste temperature, T. Hence, for a dilute system, the molar composition of the VOC in the gaseous emission, y, can be described as

$$y(T) = y^s \qquad\qquad if\, T > T^c \qquad (10.6a)$$
$$ = p^o(T)/[P^{total} - p^o(T)] \quad if\, T \le T^c, \qquad (10.6b)$$

where y^s is the supply (inlet) mole fraction of the VOC in the waste and P^{total} is the total pressure of the gas. Therefore, for a given target composition of the VOC, y^t, the VOC-recovery task is equivalent to cooling the stream to a separation-target temperature, T^*, which is calculated via the following equation:

$$y^t = p^o(T^*)/[P^{total} - p^o(T^*)]. \qquad (10.7)$$

Having converted the VOC-recovery problem into a heat-transfer task we now proceed to develop the design procedure.

10.4 Design Approach

The design procedure starts by identifying the minimum utility cost for a given heat-transfer driving force. Next, the fixed and operating costs are traded off by iterating over the driving forces until the minimum total annualized cost "TAC" is attained.

10.4.1 Minimization of External Cooling Utility

It is beneficial to develop the enthalpy expressions for the gaseous VOC-laden stream as its temperature is cooled from T^s to some arbitrary temperature, T, which is below T^c. Assuming that the latent heat of the VOC remains constant over the condensation range, the enthalpy change (e.g., kJ/kmole of VOC-free gaseous stream) can be approximated through

$$h(T) - h(T^s) = \int_{T^s}^{T} C_{p,g}(T)\,dT + \int_{T^s}^{T} y(T)C_{P,V}(T)\,dT + [y^s - y(T)]\lambda$$
$$+ \int_{T^s}^{T} [y^s - y(T)]C_{P,L}(T)\,dT. \qquad (10.8)$$

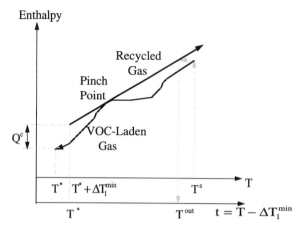

Figure 10.3 Pinch diagram for the VOC-condensation system (from Richburg and El-Halwagi, 1995. Reproduced with permission of the American Institute of Chemical Engineers. Copyright © 1995 AIChE. All rights reserved).

A similar expression (excluding the latter two terms) can be derived for the enthalpy of the recycled cold gas. A convenient way of identifying the minimum utility cost of the system is the thermal pinch diagram described in Chapter Nine. A temperature scale for the gaseous stream to be cooled, T, is related to a temperature scale for the recycled cold gas, t, by using a minimum driving force ($T = t + \Delta T_1^{min}$). Next, Eq. (10.8) is used to plot the enthalpy of the hot stream (VOC-laden stream) versus T. Similarly, one can plot the enthalpy of the cold gas recycled to the system against t. The cold stream is slid down until it touches the hot stream at the pinch point and, therefore, the minimum cooling requirement, Q^c, can be identified (Fig. 10.3). This is the load to be removed by the refrigerant in the bottom exchanger of Fig. 10.2.

For each value of β, Eq. (10.3b) is used to determine y^t which in turn is employed in Eq. (10.7) to calculate T^*. As can be seen from Fig. 10.3, for a given T^* the value of Q^c is determined graphically. In addition, the outlet temperature for the gaseous stream leaving the network, T^{out}, is identified.

10.4.2 Selection of Cooling Utilities

The next step is to screen the candidate refrigerants[1] with the objective of minimizing the utility cost. This step can be accomplished by examining the cost of

---

[1] In addition to the conventional refrigerants, new refrigerants can be synthesized to meet the technical and environmental objectives (e.g., Constantinou et al., 1995; Achenie and Duvedi, 1996; Jobak and Stephanopoulos, 1990). This area is gaining importance in light of the restrictions imposed on using halogenated compounds for refrigeration.

refrigerants that operate below T^*. Consider two refrigerants u and v whose costs ($/kJ removed) are C_u and C_v where $u > v$. It is useful to recall that the refrigerants are arranged in order of decreasing operating temperatures, hence $t_u < t_v$. If $C_u \leq C_v$, then refrigerant u is preferred over v. This rationale can be employed to compare all refrigerants below T^* with the result of identifying the one that yields the lowest cooling cost.

10.4.3 Trading Off Fixed Cost versus Operating Cost

Once the minimum utility cost has been identified, tradeoffs between operating and fixed costs must be established. This step is undertaken iteratively. For given values of minimum approach temperatures, the pinch diagram is used to obtain minimum cooling cost and outlet gas temperature. By conducting enthalpy balance around each unit, intermediate temperatures and exchanger sizing can be determined. Hence, one can evaluate the fixed cost of the system. Next, the minimum approach temperatures are altered, until the minimum TAC is identified.

10.5 Special Case: Dilute Waste Streams

In many environmental applications, the composition of the VOC is dilute enough to render the latent heat change of the stream negligible compared to its sensible heat change. In such cases, most of the cooling utility is employed to cool the carrier gas to T^*, while only a small fraction of the cooling duty is used to condense the VOC. Therefore, the hot and cold composite lines become identical in shape (linear if specific heat is assumed constant over temperature), and the pinch diagram may be approximated as shown in Fig. 10.4. On this diagram, the two composite lines touch over the overlapping temperature zone with the result of having an infinite number of pinch points.

As can be seen from Fig. 10.4, the minimum cooling utility requirement is given by:

$$Q^c = \beta G \bar{C}_{p,g} \Delta T_1^{\min},\qquad(10.9)$$

while the rest of the cooling task is fully integrated with the recycled gaseous stream. By combining Eqs. (10.3b), (10.7) and (10.9), and noting that for dilute streams $p^o(T^*) \ll P^{total}$, we get

$$Q^c = \alpha G y^s \bar{C}_{p,g} \Delta T_1^{\min}/\{y^s - [p^o(T^*)/P^{total}]\}.\qquad(10.10)$$

since $p^o(T^*)$ is a monotonically increasing function of T^*, the lower the value of T^*, the lower the Q^c. For a given refrigerant, j, and a minimum driving force $\Delta T_2^{\min}$, the lowest attainable gas temperature is $t_j + T_2^{\min}$. Hence, the optimal value of T^* can be obtained by comparing the costs of refrigerants needed to cool the

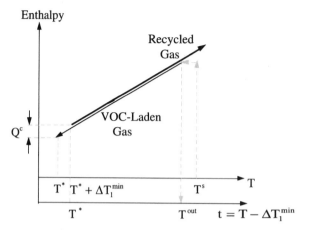

Figure 10.4 Pinch diagram for dilute VOC-condensation system.

gas to $t_j + \Delta T_2^{min}$, where $j = 1, 2, \ldots, N$. This entails searching over a finite set of at most N temperatures to identify the minimum cooling cost. Once the optimal T^* is determined, Eqs. (10.3b) and (10.7) are used to calculate the optimal β, and the optimum network is configured.

10.6 Case Study: Removal of Methyl Ethyl Ketone "MEK"

A gaseous emission contains 330 ppmv of MEK. It is desired to reduce the MEK content of this stream to 33 ppmv ($\alpha = 0.9$) prior to atmospheric discharge. The flowrate of the gas is 0.10 kmole/s, its pressure is 121,560 Pa (1.2 atm.), and its supply temperature is 300 K. Available for service are three refrigerants: NH_3, HFC134 and liquid N_2, whose respective operating temperatures are 275, 268 and 100 K and operating costs are 8, 19 and 31 \$/10^6 kJ removed, respectively. The average specific heat of the gaseous waste is taken as 30 kJ/(kmole·K). The latent heat of condensation for MEK is approximated to be 36,000 kJ/kmole over the condensation range. The vapor pressure of MEK given by (Yaws, 1994):

$$\log p^o(T) = 49.8308 - 3096.5/T - 15.184 \log T + 0.007485T$$
$$- 1.7084^* 10^{-13} T^2, \tag{10.11}$$

where $p^o(T)$ is in Pa and T is in K. In order to avoid the formation of solid MEK in the system, the lowest permissible operating temperature of the gas is 190 K, which is slightly above the freezing temperature of MEK (186.5 K).

The cost of condensers/heat exchangers (\$) is taken as 1,500*(heat transfer area in m^2)$^{0.6}$. The overall heat-transfer coefficients for the dehumidification, the integrated cooling/condensation and external cooling/condensation sections are 0.05, 0.05 and 0.10 kW/(m²·K), respectively. The fixed cost of refrigeration systems for HFC134 and NH_3 is provided by the USEPA (1991). For N_2, it is assumed that no refrigeration system is needed; instead liquid nitrogen is purchased and stored in a tank whose cost is $150,000. For all units, a linear depreciation scheme is used with 5 years of useful life period and negligible salvage value.

Solution

The supply composition of MEK is related to its partial pressure via Dalton's law:

$$y^s = \frac{p^s}{P_{total}}. \tag{10.12}$$

Hence,

$$p^s = 330 \times 10^{-6} \times 121{,}560$$
$$= 40\,\text{Pa}. \tag{10.13}$$

Similarly, the partial pressure corresponding to 33 ppmv is 4 Pa.

By combining Eqs. (10.5), (10.11) and (10.13), the temperature at which MEK starts condensing can be calculated as follows:

$$\log 40 = 49.8308 - 3096.5/T^c - 15.184 \log T^c + 0.007485 T^c$$
$$- 1.7084 * 10^{-13} T^{c2},$$

i.e.,

$$T^c = 214.5\,\text{K}. \tag{10.14}$$

Let us plot Eq. (10.11) over the condensation range as shown by Fig. 10.5. As can be seen from the figure, in order to reduce the MEK composition to 33 ppmv

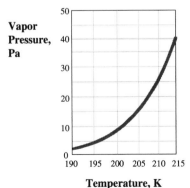

Vapor Pressure, Pa

Temperature, K

Figure 10.5 Vapor pressure vs temperature for MEK.

(partial pressure of MEK $= 4$ Pa), the temperature of the whole stream ($\beta = 1$) can be dropped to 194.5 K. Alternatively, only a fraction, β, of the stream may be passed through the condensation system. As indicated by Eq. (10.4), the bounds on β are:

$$0.900 < \beta \leq 1.000. \tag{10.15}$$

Due to the freezing limitations, the bounds are even more stringent than those given by Eq. (10.13). At the lowest permissible operating temperature of the gas (190 K), the vapor pressure of the MEK is about 2.3 Pa. According to Eq. (10.3b), the lower bound on β can be calculated as follows:

$$\frac{2.3}{121,560} = \left(1 - \frac{0.900}{\beta}\right) 330 \times 10^{-6}.$$

Hence, smallest fraction of gas to be passed through the condensation system is:

$$\beta = 0.96 \tag{10.16}$$

and Eq. (10.15) is modified to:

$$0.96 \leq \beta \leq 1.00. \tag{10.17}$$

As has been mentioned before, for each value of β, there is an equivalent value of T^* which can be determined by Eqs. (10.3b) and (10.7). Therefore, the bounds on T^* are given by

$$190 \leq T^* \leq 194.5 \tag{10.18}$$

Out of the three candidate refrigerants, only N_2 is capable of reaching this range. Therefore, N_2 is chosen as the external cooling utility. Since T^* is related to the operating temperature of the refrigerant ($t_3 = 100$ K) by

$$T^* = 100 + \Delta T_2^{min}, \tag{10.19}$$

which can be combined with Eq. (10.18) to give:

$$90.0 \leq \Delta T_2^{min} \leq 94.5. \tag{10.20}$$

Due to the dilute nature of the stream, we may assume that the latent heat of condensation for MEK is much smaller than the sensible heat removed from the gas. Therefore, we can apply the procedure presented in Section 10.5. Once a value is selected for ΔT_2^{min}, Eq. (10.19) can be used to determine T^* and Eqs. (10.3b) and (10.7) can be employed to calculate the value of β. Since the bounds on β (and consequently on ΔT_2^{min}) are tight, we will iterate over two values of β; 0.96 and 1.00 ($\Delta T_2^{min} = 90.0$ and 94.5 K). The other iterative variable is ΔT_1^{min}. Both variables are used to trade off fixed versus operating costs. As an illustration, consider the following iteration: $\Delta T_1^{min} = 5$ K and

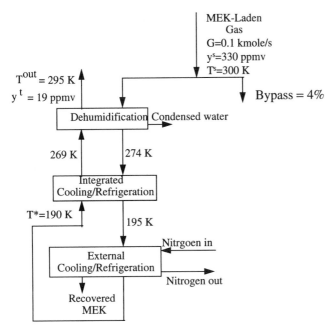

Figure 10.6 The network configuration for the MEK removal example (at $\Delta T_1^{min} = 5$ K and $\Delta T_2^{min} = 90$ K).

$\Delta T_2^{min} = 90$ K. According to Eq. (10.19), $T^* = 190$ K, which corresponds to $\beta = 0.96$ (Eq. 10.16). Using heat balance, the network can be configured as shown in Fig. 10.6. Let us now proceed to size and estimate the cost of the system.

The minimum cooling utility is calculated using Eq. (10.9),

$$Q^c = 0.96 \times 0.1 \times 30 \times 5$$
$$= 14.4 \, \text{kW},\qquad(10.21)$$

which corresponds to the following utility cost:

$$\text{Annual operating cost of refrigerant} = 14.4\frac{kJ}{s} \times 3600\frac{s}{hr} \times 8760\frac{hr}{yr} \times 31\frac{\$}{10^6 \, kJ}$$
$$= \$14,100/\text{yr}.\qquad(10.22)$$

The exchangers can be sized as follows:

$$A = \frac{Q}{U \cdot \Delta T_{lm}},\qquad(10.23)$$

where A is the heat-transfer area, Q is the heat transferred, U is the overall heat transfer coefficient and ΔT_{lm} is the logarithmic mean temperature difference. For instance, the area of the first exchanger can be estimated as follows:

$$A_1 = \frac{0.96 \times 0.1 \times 30 \times (300 - 274)}{0.05 \times 5}$$

$$= 300 \text{ m}^2. \tag{10.24}$$

whose fixed cost is $1,500\ (300^{0.6}) = \$46,000$. Similarly, the other exchangers can be sized and the total fixed cost of the second and third exchangers is calculated to be \$89,000 and 2,000, respectively. When the cost of the nitrogen tank is added and the fixed cost is depreciated over five years, the annualized fixed cost of the system is \$57,400/yr. Combining this result with Eq. (10.22), we get a total annualized cost of \$71,500/yr. The same procedure is repeated for several values of $\Delta T_1^{\min}$ and $\Delta T_2^{\min}$. The results are plotted in Fig. 10.7 for $\beta = 0.96$ (which corresponds to $\Delta T_2^{\min} = 90$ K). As can be seen from the figure, the optimal value of $\Delta T_1^{\min}$ is about 5 K, and the minimum total annualized cost is approximately \$71,500/yr. Hence, the optimal system is the configuration illustrated by Fig. 10.6.

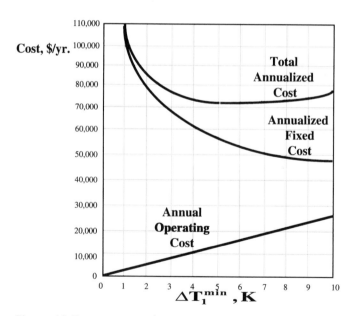

Figure 10.7 Cost versus $\Delta T_1^{\min}$ for the MEK example (at $\beta = 0.96$).

10.7 Effect of Pressure

Pressure is an important factor in condensation. In general, the task of condensing VOCs entails the use of an energy-induced separation network "EISEN" in which pressure and cooling energies are employed to induced condensation (Dunn *et al.*, 1995). Let us revisit the MEK example and assign a target composition of 11 ppmv MEK. With the gaseous stream operating at 1.2 atm., MEK will freeze before reaching this target. Alternatively, the stream can be pressurized to induce additional condensation. Using Eq. (10.7), at the lowest permissible gas temperature (190 K) we get

$$11 \times 10^{-6} = \frac{2.26}{P^{total}},$$

i.e.,

$$P^{total} = 205,456 \, \text{Pa (or 2.0 atm)}. \tag{10.25}$$

Hence, the target composition can be achieved by compressing the gaseous waste to 2.0 atm. In some cases, following condensation the stream may be throttled using a turbine which recovers the pressure energy and may induce additional condensation. In such cases, the costs of cooling, compression and depressurization should be traded off (e.g., see Dunn *et al.*, 1995, and homework problem 10.2).

Problems

10.1 In a polyvinyl chloride plant (Richburg and El-Halwagi, 1995), a continuous air emission leaves a dryer stack. The primary pollutant in this stream is VC. The flowrate of the gas is 0.20 kmole/s, its supply temperature is 338 K and contains 0.5 mol/mol% of VC. It is desired to recover 90% of the VC in the gaseous waste. Available for service are four refrigerants: SO_2, HFC134, NH_3 and N_2 whose respective operating temperatures are 280, 265, 245 and 140 K and operating costs are 12, 10, 13 and 16 \$/10^6 kJ removed, respectively. The average specific heats of the air, VC vapor and VC liquid are taken as 29, 50 and 85 kJ/(kmole·K), respectively. The latent heat of condensation for VC is assumed to be 24,000 kJ/kmole over the operating range. The vapor pressure of VC is given by (Yaws, 1994):

$$\log p^o(T) = 52.9654 - 2.5016 \times 10^3/T - 17.914 \log T + 0.0108 \, T$$
$$- 4.531 \times 10^{-14} T^2, \tag{10.26}$$

where $p^o(T)$ is in mm Hg and T is in K. The cost of condensers/heat exchangers (\$) is taken as 1,500*(heat transfer area in m^2)$^{0.6}$. The overall heat-transfer coefficients for the dehumidification, the integrated cooling/condensation and external cooling/condensation sections are 0.05, 0.05 and 0.10 kW/(m^2· K), respectively. The fixed cost of refrigeration systems for SO_2, HFC134 and NH_3 is provided elsewhere (USEPA, 1991). For N_2, it is assumed that no refrigeration system is needed; instead, liquid nitrogen is purchased and stored in a tank whose cost is $150,000. For all units, a linear depreciation scheme is used with 10 years of useful life period and negligible salvage value. Design a condensation system which features the minimum total annualized cost.

10.2 A gaseous emission has a flowrate of 0.02 kmole/s and contains 0.014 mole fraction of vinyl chloride. The supply temperature of the stream is 338 K. It is desired to recover 80% of the vinyl chloride using a combination of pressurization and cooling. Available for service are two refrigerants: NH_3 and N_2. Thermodynamic and economic data are provided by problem 10.1 and by Dunn *et al.* (1995). Design a cost-effective energy-induced separation system.

10.3 The techniques presented in this chapter can be extended to streams with multi-component VOCs. Consider a magnetic-tape manufacturing plant which disposes of a VOC-laden gaseous emission (Dunn and El-Halwagi, 1994a). This stream contains four main VOC's tetrahydrofuran "THF", methyl ethyl ketone "MEK", toluene and cyclohex-anone. The composition of these compounds in the gaseous emission are 723, 865, 114 and 849 ppmv, respectively. Due to environmental regulations, the organic loading of these compounds should be reduced by 95 mol.% via condensation. Six refrigerants are being considered; HCFC142b, HFC134a, HCFC22, SO_2, NH_3 and N_2. Synthesize a minimum-cost condensation system. (**Hint:** see Dunn and El-Halwagi, 1994a).

Symbols

C_j	cost of the jth refrigerant, $/kJ removed
$C_{p,g}$	specific heat of the VOC-free gas, kJ/(kmole·K)
$\bar{C}_{p,g}$	average specific heat of VOC-free gas, kJ/(kmole·K)
$C_{P,L}$	specific heat of the liquid VOC, kJ/(kmole·K)
$C_{P,V}$	specific heat of the vapor VOC, kJ/(kmole·K)
G	flowrate of VOC-free gaseous stream, kmole/s
h	specific enthalpy of gaseous stream, kJ/kmole of VOC-free gaseous stream
j	index for refrigerants
N	total number of potential refrigerants
$p^o(T)$	vapor pressure of the VOC at T, mm Hg
p^{total}	pressure of gaseous stream, mm Hg
Q	rate of heat transfer, kW
Q^c	minimum cooling requirement, kW
t	temperature of cold gas to be heated, K
t_j	operating temperature of the jth refrigerant, K
T	temperature of gaseous stream being cooled, K
T^c	dew temperature for the VOC, K
T^s	supply (inlet) temperature of gaseous stream, K
T^*	separation-target temperature to which the gaseous stream has to be cooled, K

y^s supply composition of VOC, kmole VOC/kmole VOC-free gaseous stream
y^t target composition of VOC, kmole VOC/kmole VOC-free gaseous stream

Greek

α recovery fraction of VOC in gaseous stream, kmole VOC recovered/kmole
         VOC in gaseous waste
β fraction of gaseous stream to be passed through the condensation system
ΔT_1^{min} minimum driving force for the dehumidification and integrated cool-
         ing/condensation blocks, K
ΔT_2^{min} minimum driving force for the external cooling/condensation block, K
λ latent heat of condensation for VOC, kJ/kmole

References

Achenie, L., and Duvedi, A. (1996). "A Mixed Integer Nonlinear Programming Modle for the Design of Refrigerant Mixtures," *AIChE Annual Meeting*, San Francisco.

Constantinou, L., Jaksland, C., Bagherpour, K., Gani, R., and Bogle, I. D. L. (1995). Application of the group contribution approach to tackle environmentally related problems. *AIChE Symp. Ser.*, **90**(303), 105–106, NY: AIChE.

Dunn, R. F., and El-Halwagi, M. M. (1994a). Optimal design of multicomponent VOC condensation systems, *J. Haz. Mat.*, **38**, 187–206.

Dunn, R. F., and El-Halwagi, M. M. (1994b). Selection of optimal VOC condensation systems. *J. Waste Management*, **14**(2), 103–113.

Dunn, R. F., Zhu, M., Srinivas, B. K., and El-Halwagi, M. M. (1995). Optimal design of energy induced separation networks for VOC recovery. *AIChE Symp. Ser.*, **90**(303), 74–85, NY: AIChE.

El-Halwagi, M. M., Srinivas, B. K., and Dunn, R. F. (1995). Synthesis of optimal heat induced separation networks. *Chem. Eng. Sci.*, **50**(1), 81–97.

Joback, K. G., and Stephanopoulos, G. (1990). Designing molecules possessing desired physical property values. In "Foundations of Computer Aided Process Design 'FOCAPD' III" (J. J. Siirola, I. Grossmann, and G. Stephanopoulos, eds.), pp. 363–387, CACHE/Elsevier, New York.

Richburg, A., and El-Halwagi, M. M. (1995). "A graphical approach to the optimal design of heat-induced separation networks for VOC recovery," *AIChE Symp. Ser.*, **91**(304), 256–259, NY: AIChE.

U. S. EPA (1991). "Control Technologies for Hazardous Air Pollutants," EPA/625/6-91/014, 4.55–4.64. Washington, DC.

Yaws, C. L. (1994). "Handbook of Vapor Pressure," Vol. 1, p. 342. Gulf Pub. Co., Houston. TX.

CHAPTER ELEVEN

Design of
Membrane-Separation Systems

Recently, membrane processes have gained a growing level of applicability in industry. Membrane systems have several advantages. In addition to their high selectivity (that can provide concentrations as low as parts per billion), low energy consumption and moderate cost, they are compact and modular. Therefore, membrane units can be readily added to existing plants. This chapter provides an overview of the use of membrane-separation systems for environmental applications. First, pressure-driven membrane technologies will be categorized. Then, modeling and design techniques will be discussed for reverse osmosis as a representative membrane technology. For more details on the subject, the reader is referred to Crabtree *et al.* (1997), El-Halwagi *et al.* (1996), Crespo and Boddeker (1994), Matsuura (1993), El-Halwagi (1992), Mulder (1991), Bitter (1991), Baker *et al.* (1991), Rautenbach and Albrecht (1989), Parekh (1988), Sourirajan and Matsuura (1985) and Belfort (1984).

11.1 Classification of Membrane Separations

The most common membrane systems are driven by pressure. The essence of a pressure-driven membrane process is to selectively permeate one or more species through the membrane. The stream retained at the high pressure side is called the retentate while that transported to the low pressure side is denoted by the permeate (Fig. 11.1). Pressure-driven membrane systems include microfiltration, ultrafiltration, reverse osmosis, pervaporation and gas/vapor permeation. Table 11.1 summarizes the main features and applications of these systems.

Table 11.1
Features of Pressure-Driven Membrane Systems for Environmental Applications

Process	Retentate	Permeate	Common range of feed pressure (atm)	Membrane type	Typical applications/ species
Microfiltration	Liquid	Liquid	1.5–7	Porous	Organic and metal suspensions, oil/water emulsions
Ultrafiltration	Liquid	Liquid	2–15	Porous	Oil/water emulsions, pesticides, herbicides bivalent ions
Reverse osmosis (hyperfiltration or nanofiltration)	Liquid	Liquid	10–70	Porous/ nonporous	Desalination, salts, organics, ions heavy metals
Pervaporation	Liquid	Vapor	1.5–60 (permeate is under vacuum)	Nonporous	Volatile organic compounds
Vapor permeation	Vapor	Vapor	Less than saturation pressure of feed (permeate is under vacuum)	Nonporous	Volatile organic compounds
Gas permeation	Gas	Gas	3.0–55	Nonporous	He, H_2, NO_x, CO, CO_2, hydrocarbons chlorinated hydrocarbons

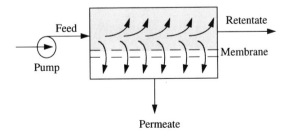

Figure 11.1 Schematic representation of a membrane separation unit.

Among the technologies listed in Table 11.1, reverse osmosis RO has gained significant commercial acceptance. Therefore, the remainder of this chapter will discuss RO systems as representative of membrane-separation technologies.

11.2 Reverse Osmosis Systems

Over the past three decades, there has been a growing industrial interest in using reverse osmosis for several objectives such as water purification and demineraliza-tion as well as environmental applications (e.g., Comb, 1994; Rorech and Bond, 1993, El-Halwagi, 1992). The first step in designing the system is to understand the operating principles and modeling of RO modules.

11.2.1 Operating Principles

When two solutions with different solute (pollutant) concentrations are separated by a semi-permeable membrane, a chemical-potential difference arises across the membrane. This difference causes the carrier (typically water) to permeate through the membrane from the low-concentration (high chemical potential) side to the high-concentration (low chemical potential) side. This flow is referred to as the *osmotic flow* (Fig. 11.2a). Osmotic flow causes an increase in the pressure head of the high-concentration side. This flow continues until the pressure difference across the membrane balances the difference in the chemical potential across the membrane and, therefore, reaching equilibrium. The pressure head at equilibrium is termed *osmotic pressure* (Fig. 11.2b). The osmotic phenomenon can be reversed by applying a pressure greater than the osmotic pressure on the high-concentration side. Therefore, reverse osmosis is based on the phenomenon that takes place when an external pressure (larger than the osmotic pressure) is applied to a solution (e.g.,

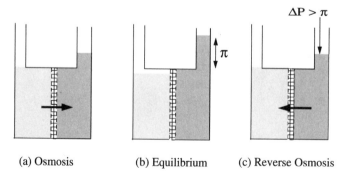

(a) Osmosis   (b) Equilibrium   (c) Reverse Osmosis

Figure 11.2 Direct and reverse osmosis.

aqueous waste) causing the carrier (e.g., water) to preferentially permeate, while the solute (e.g., pollutant) is rejected and left in the retentate (Fig. 11.2c). Osmotic pressures can be evaluated using chemical potential calculations. However, for conceptual design purposes, approximate expressions can be employed.[1]

Typically RO systems are preceded by pretreatment units to remove suspended solids/colloidal matter and add chemicals that control biological growth and reduce scaling. Membranes are typically made of synthetic polymers coated on a backing (skin). Examples of membrane materials include polyamides, cellulose acetate and sulfonated polysulfone.

There are three main configurations of RO units, hollow fiber, tubular and spiral wound. In particular, hollow-fiber reverse-osmosis "HFRO" systems have received considerable industrial attention. Among the different configurations of semipermeable membranes used for reverse osmosis, the hollow-fiber modules have a number of distinguishing characteristics such as the large surface-to-volume ratio, the self-supporting strength of fibers and the negligible concentration polarization near the membrane surface. The next section focuses on the modeling of HFRO units.

11.2.2 Modeling of HFRO Units

A hollow-fiber reverse-osmosis module consists of a shell which houses the hollow fibers (Fig. 11.3). The fibers are grouped together in a bundle with one end sealed and the other open to the atmosphere. The open ends of the fibers are potted into an epoxy sealing head plate after which the permeate is collected. The pressurized feed solution (denoted by the shell side fluid) flows radially from a central porous tubular distributor. As the feed solution flows around the outer side of the fibers toward the shell perimeter, the permeate solution penetrates through the fiber wall into the bore side by virtue of reverse osmosis. The permeate is collected at the open ends of the fibers. The reject solution is collected at the porous wall of the shell.

In modeling an RO unit, two aspects should be considered; membrane transport equations and hydrodynamic modeling of the RO module. The membrane transport equations represent the phenomena (water permeation, solute flux, etc.) taking place at the membrane surface. On the other hand, the hydrodynamic model deals with the macroscopic transport of the various species along with the momentum and energy associated with them. In recent years, a number of mathematical

---

[1]For example, in the case of dilute solutions, the van't Hoff's equation may be used to predict the osmotic pressure ($\pi = CRT$) where π is the osmotic pressure of the solution, C is the molar concentration of the solute, R is the universal gas constant and T is the absolute temperature. For dissociating solutes, the concentration is that of the total ions. For example, NaCl dissociates in water into two ions; Na^+ and Cl^-. Therefore, the total molar concentration of ions is *twice* the molar concentration of NaCl. A useful rule of thumb for predicting osmotic pressure of aqueous solutions is 0.01 psi/ppm of solute (Weber, 1972).

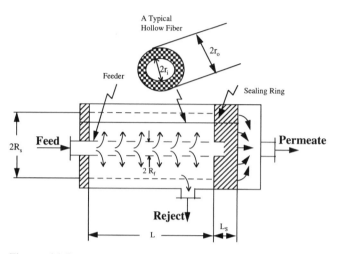

Figure 11.3 Schematic representation of an HFRO module.

models have been devised to simulate the hydrodynamic performance of HFRO modules. In these models, two approaches have been generally adopted to estimate the pressure variation through the shell side. In the first approach, the shell-side pressure is assumed to be constant (Gupta, 1987; Soltanieh and Gill, 1982, 1984; Ohya *et al.*, 1977). The second approach treats the fiber bundle as a porous medium and describes the flow through this medium by Darcy's equation with an arbitrary empirical constant (Kabadi *et al.*, 1979; Hermans, 1978; Orofino, 1977; Dandavati *et al.*, 1975). The value of the empirical constant is strictly applicable to the geometry and operating conditions for which it has been evaluated. As a result of the pressure drop inside the fibers along the module, the pressure difference across the fiber wall may vary significantly and, thus, the permeation rate may change considerably along the fiber length. Therefore, an axial component of the shell-side flow arises in addition to the radial component. El-Halwagi *et al.* (1996) developed a two-dimensional model that captures the radial and axial flows within an HFRO module.

Assuming that the densities of the feed, the permeate and the reject are the same, one can write the following overall material balance around the module:

$$q_F = q_P + q_R, \tag{11.1}$$

where q_F, q_P and q_R are the volumetric flowrates of the feed, the permeate and the reject, respectively, per module. A component material balance on the solute is given by

$$q_F C_F = q_P C_P + q_R C_R. \tag{11.2}$$

In addition to material balance, two transport equations can be used to predict the flux of water and solute. For instance, the following simplified model can be used (Dandavati *et al.*, 1975; Evangelista, 1986).

Water flux

$$N_{water} = A\left(\Delta P - \frac{\pi_F}{C_F}C_S\right)\gamma \tag{11.3}$$

where N_{water} is water flux, ΔP is pressure difference across the membrane, π_F, is osmotic pressure of feed, C_F is solute concentration in the feed, C_S is average solute concentration in the shell side and γ is given by

$$\gamma = \frac{\eta}{1 + \dfrac{16A\mu r_o L L_s \eta}{1.0133 \times 10^5 r_i^4}}, \quad where$$

$$\eta = \frac{\tanh\theta}{\theta} \quad and \tag{11.4}$$

$$\theta = \left(\frac{16A\mu r_o}{1.0133 \times 10^5 r_i^2}\right)^{\frac{1}{2}} \frac{L}{r_i}.$$

In many cases, Eq. (11.3) may be further simplified by assuming linear shell-side concentration and pressure profiles, i.e.,

$$\Delta P \approx \frac{P_F + P_R}{2} - P_P$$

$$= \frac{P_F + P_F - \text{shell side pressure drop per module}}{2} - P_P$$

$$= P_F - \left(\frac{\text{shell side pressure drop per module}}{2} + P_P\right) \tag{11.5}$$

where P_F, P_R and P_P are pressures of feed, reject and permeate, respectively. Similarly,

$$C_S \approx \frac{C_F + C_R}{2}. \tag{11.6}$$

Solute flux

$$N_{solute} = \left(\frac{D_{2M}}{K\delta}\right)C_S$$

$$\approx \left(\frac{D_{2M}}{K\delta}\right)\left(\frac{C_F + C_R}{2}\right). \tag{11.7}$$

Permeate flowrate The volumetric flowrate of the permeate per module can be obtained from the water flux as follows:

$$q_P = S_m N_{water}, \tag{11.8}$$

where S_m is the surface area of the hollow fibers per module. By combining Eqs. (11.3), (11.6) and (11.8), we get

$$N_{water} = A \left[\Delta P - \frac{\pi_F}{2} \left(1 + \frac{C_R}{C_F} \right) \right] \gamma \tag{11.9a}$$

and

$$q_p = S_m A \left[\Delta P - \frac{\pi_F}{2} \left(1 + \frac{C_R}{C_F} \right) \right] \gamma. \tag{11.9b}$$

Permeate concentration The solute concentration in the permeate may be approximated by the ratio of the solute to water fluxes, i.e.,

$$C_P \approx \frac{N_{solute}}{N_{water}} \tag{11.10}$$

Equations (11.2)–(11.10) provide a complete model for describing the performance of an HFRO module. For a given set of q_F, C_F, P_F and P_P (typically atmospheric), along with the values of the module physical properties, one can predict q_R, q_P, C_R and C_P. A particularly useful solution scheme is in the case of highly rejecting membranes. In this case, most of the solute is retained in the reject. Therefore, Eq. (11.2) can be simplified to

$$q_F C_F \approx (q_F - q_P) C_R. \tag{11.11}$$

Combining Eqs. (11.9b) and (11.11), we obtain

$$q_F C_F = \left\{ q_F - S_m A \left[\Delta P - \frac{\pi_F}{2} \left(1 + \frac{C_R}{C_F} \right) \right] \gamma \right\} C_R.$$

Hence,

$$S_m A \frac{\pi_F}{2 C_F} \gamma C_R^2 + \left[q_F - S_m A \left(\Delta P - \frac{\pi_F}{2} \right) \gamma \right] C_R - q_F C_F = 0, \tag{11.12}$$

which is a quadratic equation that can be solved for C_R. Having identified the value of C_R, Eqs. (11.6), (11.7), (11.9a), (11.9b) and (11.10) are used to determine the values of C_S, N_{solute}, N_{water}, q_P and C_P, respectively. It is worth pointing out that the validity of Eq. (11.11) can be confirmed when $q_p C_P \ll q_F C_F$ (highly rejecting membranes).

The foregoing equations assume that membrane performance is time independent. In some cases, a noticeable reduction in permeability occurs over time primarily due to membrane fouling. In such cases, design and operational provisions are used to maintain a steady performance of the system (Zhu et al., 1997).

Example 11.1: Modeling the Removal of an Inorganic Salt from Water

Let us consider the following case of removing an inorganic salt from an aqueous stream. It is desired to reduce the salt content of a 26 m^3/hr water stream (Q_F) whose feed concentration, C_F, of 0.035 kmol/m^3 (approximately 2,000 ppm). The feed osmotic pressure (π_F) is 1.57 atm. A 30 atm (P_F) booster pump is used to pressurize the feed. Sixteen hollow fiber modules are to be employed for separation. The modules are configured in parallel with the feed distributed equally among the units. The following properties are available for the HFRO modules:

$$\text{Permeability}(A) = 5.573 \times 10^{-8} \frac{m}{s \cdot atm}$$

$$\text{Salt flux constant}\left(\frac{D_{2M}}{K\delta}\right) = 1.82 \times 10^{-8} m/s$$

$$\text{Viscosity}(\mu) = 10^{-3} \frac{kg}{ms}$$

$$\text{Outside radius of fibers }(r_o) = 42 \times 10^{-6} \text{ m}$$

$$\text{Inside radius of fibers }(r_i) = 21 \times 10^{-6} \text{ m}$$

$$\text{Fiber length }(L) = 0.75 \text{ m}$$

$$\text{Seal length }(L_S) = 0.075 \text{ m}$$

$$\text{Membrane area per module }(S_m) = 180 \text{ m}^2$$

$$\text{Pressure drop per module} = 0.4 \text{ atm.}$$

Estimate the values of the permeate flowrate and concentration.

Solution

Since the 16 modules are in parallel, they will behave the same. Therefore, we can model one module and deduce the result for the rest of the modules. The feed flowrate per module is given by:

$$q_F = \frac{Q_F}{n}$$

$$= \frac{26.0}{(16)(3600)}$$

$$= 4.5 \times 10^{-4} \frac{m^3}{module \cdot s}. \tag{11.13}$$

Pressure drop across the membrane can be calculated using Eq. (11.5):

$$\Delta P = 30.0 - \left(\frac{0.4}{2} + 1.0\right)$$

$$= 28.8 \; atm.$$

Module Properties (Eq. 11.4):

$$\theta = \left[\frac{(16)(5.5573 \times 10^{-8})(0.001)(42 \times 10^{-6})}{(1.0133 \times 10^5)(21 \times 10^{-6})^2}\right]^{\frac{1}{2}} \left(\frac{0.75}{21 \times 10^{-6}}\right)$$

$$= 1.0339$$

$$\eta = \frac{\tanh(1.0339)}{1.0339}$$

$$= 0.75$$

$$\gamma = \frac{0.75}{1 + \dfrac{(16)(0.5573 \times 10^{-7})(0.001)(42 \times 10^{-6})(0.75)(0.075)(0.75)}{(1.0133 \times 10^5)(21 \times 10^{-6})^4}}$$

$$= 0.6943$$

By substituting the values of the various parameters in Eq. (11.12), we get

$$1.562 \times 10^{-4} C_R^2 + 2.549 \times 10^{-4} C_R - 1.575 \times 10^{-5} = 0.$$

Solving for C_R, we get

$$C_R = 0.0596 \; kmol/m^3 \quad \text{(the other root is rejected for being negative)}$$

Using Eq. (11.6), we have

$$C_s \approx \frac{0.0350 + 0.0596}{2}$$

$$= 0.0473 \; kmol/m^3.$$

The solute flux can be determined from Eq. (11.7):

$$N_{solute} \approx (1.82 \times 10^{-8})(0.0473)$$

$$= 8.61 \times 10^{-10} \frac{kmol}{m^2 s}$$

The water flux can be calculated from Eq. (11.9a) as follows:

$$N_{water} = 5.573 \times 10^{-8}\left[28.8 - \frac{1.57}{2}\left(1 + \frac{0.0596}{0.035}\right)\right]0.6943$$

$$= 1.032 \times 10^{-6} \text{ m/s,}$$

and from Eq. (11.9b), the volumetric flowrate of the permeate per module

$$q_p = 180 \times 1.032 \times 10^{-6}$$

$$= 1.86 \times 10^{-4}\frac{\text{m}^3}{\text{module} \cdot \text{s}}.$$

Hence, the volumetric flowrate of the permeate from the 16 modules, Q_P, is 10.7 m^3/hr (2.97×10^{-3} m^3/s).

The solute concentration in the permeate may be obtained from Eq. (11.10) as follows

$$C_P \approx \frac{8.61 \times 10^{-10}}{1.032 \times 10^{-6}}$$

$$= 8.34 \times 10^{-4} \text{ kmol/m}^3 \quad (48 \text{ ppm}).$$

Example 11.2: Estimation of Transport Parameters

In a permeation experiment, an HFRO module with a membrane area of 200 m^2 is used to remove a nickel salt from an electroplating wastewater. The feed to the module has a flowrate of 5×10^{-4} m^3/s, a nickel-salt composition of 4,000 ppm and an osmotic pressure of 2.5 atm. The average pressure difference across the membrane is 28 atm. The permeate is collected at atmospheric pressure. The results of the experiment indicate that the water recovery[2] is 80% while the solute rejection[3] is 95%. Evaluate the transport parameters $A\gamma$ and $(D_{2M}/K\delta)$.

Solution

Let us first evaluate the composition and flowrate of the permeate using solute rejection and permeate recovery:

$$C_p = (1 - 0.95)4,000$$

$$= 200 \text{ ppm}$$

---

[2]Water recovery (or conversion) is the ratio of the permeate flowrate to the feed flowrate $\times$ 100%.
[3]Solute rejection is defined as $(1 - C_p/C_F) \times 100\%$.

and

$$q_p = 0.8 \times 5 \times 10^{-4}$$
$$= 4 \times 10^{-4} \text{ m}^3/\text{s}.$$

Therefore, according to Eq. (11.8):

$$N_{\text{water}} = \frac{4 \times 10^{-4}}{200}$$
$$= 2 \times 10^{-6} \text{ m/s}.$$

The reject flowrate can be evaluated via material balance:

$$q_R = 5 \times 10^{-4} - 4 \times 10^{-4}$$
$$= 1 \times 10^{-4} \text{ m}^3/\text{s}.$$

Using the component material balance, we get

$$C_R = \frac{5 \times 10^{-4} \times 4000 - 4 \times 10^{-4} \times 200}{1 \times 10^{-4}}$$
$$= 19,200 \text{ ppm}.$$

The product of the transport parameters A and γ can be calculated from Eq. (11.9a):

$$2 \times 10^{-6} = A\gamma \left[28 - \frac{2.5}{2.0} \left(1 + \frac{19200}{4000} \right) \right],$$

i.e.,

$$A\gamma = 9.64 \times 10^{-8} \frac{m}{s \cdot atm}.$$

Therefore, for this system the water flux equation (11.9a) is:

$$N_{water} = 9.64 \times 10^{-8} \left[\Delta P - \pi_F \left(1 + \frac{C_R}{C_F} \right) \right].$$

We now turn our attention to the solute flux. Using Eq. (11.10), we get

$$N_{\text{solute}} = 200 \times 2 \times 10^{-6}$$
$$= 4 \times 10^{-4} \frac{\text{ppm} \cdot \text{m}}{\text{s}}.$$

By employing Eq. (11.7), we can evaluate the salt flux constant:

$$\left(\frac{D_{2M}}{K\delta} \right) = \frac{2 \times 4 \times 10^{-4}}{4000 + 19200}$$
$$= 3.45 \times 10^{-8} \text{ m/s}.$$

11.3 Designing Systems of Multiple Reverse Osmosis Modules

In most industrial applications, it is rare that a single RO module can be used to address the separation task. Instead, a *reverse-osmosis network (RON)* is employed. A RON is composed of multiple RO modules, pumps and turbines. The following sections describe the problem of synthesizing a system of RO modules and a systematic procedure for designing an optimal RON. Once a RON is synthesized, it can be incorporated with a mass integration framework (see Problem 11.6).

11.3.1 Synthesis of RONs: Problem Statement

The task of synthesizing an optimal RON can be stated as follows: For a given feed flowrate, Q_F, and a feed concentration, C_F, it is desired to synthesize a minimum cost system of reverse osmosis modules, booster pumps and energy-recovery turbines that can separate the feed into two streams; an environmentally acceptable permeate and a retentate (reject) stream in which the undesired species is concentrated. The permeate stream must meet two requirements:

1. The permeate flowrate should be no less than a given flowrate, i.e.,

$$Q_P \geq Q_P^{\min}. \tag{11.14}$$

2. The concentration of the undesirable species in the permeate should not exceed a certain limit (typically the environmental regulation):

$$C_P \leq C_P^{\max}. \tag{11.15}$$

The flowrate per module is typically bounded by manufacturer's constraints:

$$q_F^{\min} \leq q_F \leq q_F^{\max}. \tag{11.16}$$

Figure 11.4 shows a schematic representation of the RON problem.

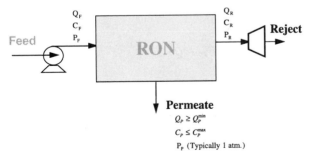

Figure 11.4 The RON synthesis problem.

11.3.2 A Shortcut Method for Synthesis of RONs

The problem of designing a RON entails the identification of membrane types, sizes, number and arrangement. In addition, the designer has to determine the optimal operating conditions and the type, number and size of any pumps and energy-recovery devices. In order to understand the basic principles of synthesizing an optimal RON, let us consider the class of problems for which one stage of parallel RO modules is used.[4] A booster pump is first used to raise the pressure to its optimal level. The feed is distributed among a number, n, of parallel modules. The reject is collected as a retentate stream which is fed to an energy recovery turbines (if the value of recovered energy is higher than the cost of recovering it). The permeate streams are also gathered and constitute the environmentally-acceptable stream. The following design and operating variables are to be optimized]:

1. Feed pressure, P_F, and consequently size of booster pump.
2. Number of parallel modules, n.
3. Size of energy-recovery turbine.

It is instructive to consider the equations specifying the system performance. Overall material balance for the RON:

$$Q_F = Q_P + Q_R, \tag{11.17}$$

where Q_F, Q_P and Q_R are the volumetric flowrates of the feed, the permeate and the reject, respectively, for the whole RON.
System and module flowrates:

$$Q_F = nq_F \tag{11.18}$$

$$Q_P = nq_P \tag{11.19}$$

$$Q_R = nq_R \tag{11.20}$$

Solute material balance around the RON:

$$Q_F C_F = Q_P C_P + Q_R C_R. \tag{11.21}$$

Water flux:

$$N_{water} = A\left(\Delta P - \frac{\pi_F}{C_F}C_S\right)\gamma. \tag{11.3}$$

[4]For the cases of multiple stages of RO modules, the reader is referred to Evangelista (1985, 1986), El-Halwagi (1992, 1993) and El-Halwagi and El-Halwagi (1992).

Pressure drop across the membrane

$$\Delta P \approx \frac{P_F + P_R}{2} - P_P$$

$$= \frac{P_F + P_F - \text{shell side pressure drop per module}}{2} - P_P \qquad (11.5)$$

$$= P_F - \left(\frac{\text{shell side pressure drop per module}}{2} + P_P \right)$$

Average shell-side pressure:

$$C_S \approx \frac{C_F + C_R}{2}. \qquad (11.6)$$

Solute flux:

$$N_{\text{solute}} = \left(\frac{D_{2M}}{K\delta} \right) C_S$$

$$\approx \left(\frac{D_{2M}}{K\delta} \right) \left(\frac{C_F + C_R}{2} \right) \qquad (11.7)$$

Permeate flowrate per module:

$$q_P = S_m N_{\text{water}}. \qquad (11.8)$$

Solute concentration in the permeate:

$$C_P \approx \frac{N_{solute}}{N_{water}}. \qquad (11.10)$$

These 11 equations describe the performance of the single-stage RON. For a given Q_F, C_F and Q_P,[5] there are 12 unknown variables (C_P, C_R, C_S, n, N_{water}, N_{solute}, P_F, q_F, q_P, q_R, Q_R, and ΔP). Therefore, by fixing one variable the whole system is completely specified. Hence, the system can be optimized by iteratively varying one variable and evaluating the cost at each iteration. For instance, the feed pressure can be varied total annualized cost of the system can be plotted versus pressure to locate the optimum feed pressure. Nonetheless, in order to avoid the simultaneous solution of 11 equations, it is recommended to vary C_p iteratively. The following iterative procedure can be used to optimize the system:

1. Setting the flowrate of the permeate to be Q_p^{min}, we can calculate the reject flowrate by rearranging the material balance (Eq. 11.17) as follows:

$$Q_R = Q_F - Q_P. \qquad (11.17)$$

[5]In the problem statement, it is required to meet a minimum demand on permeate flowrate. Typically, it is possible to provide an actual flowrate that is very close to the minimum required flowrate by adjusting the number of modules.

2. Select an iterative value of C_P which satisfies Eq. (11.15). Therefore, one can calculate the reject concentration by rearranging the component material balance on the solute (Eq. 11.21), i.e.,

$$C_R = \frac{Q_F C_F - Q_P C_P}{Q_R}. \tag{11.21}$$

3. Calculate the average shell-side concentration according to Eq. (11.6):

$$C_S \approx \frac{C_F + C_R}{2}. \tag{11.6}$$

4. Calculate the salt flux according to Eq. (11.7):

$$N_{solute} = \left(\frac{D_{2M}}{K\delta}\right) C_S \tag{11.7}$$

5. Evaluate the water flux using Eq. (11.10):

$$N_{water} \approx \frac{N_{solute}}{C_P}. \tag{11.10}$$

6. Determine the pressure difference across the membrane, ΔP, by solving Eq. (11.3) where the only unknown is ΔP:

$$N_{water} = A\left(\Delta P - \frac{\pi_F}{C_F} C_s\right)\gamma. \tag{11.3}$$

7. Calculate the feed pressure using Eq. (11.5):

$$P_F \approx \Delta P + \left(\frac{\text{shell side pressure drop per module}}{2} + P_P\right), \tag{11.5}$$

where P_P is typically one atmosphere.

8. Evaluate the permeate flowrate per module using Eq. (11.8):

$$q_P = S_m N_{water}. \tag{11.8}$$

9. In order to determine the number of parallel modules, n, calculate Q_p/q_p then set n equal to the next larger integer.

10. Now that the values of P_F and n have been determined, we can model the system as described by Section 11.2. With the proper cost functions, the total annualized cost "TAC" of the system can be evaluated for this iteration. Next, a new value of C_P is selected and steps 2–10 are repeated. The TAC of the system can be plotted vs. C_P (or any of the 10 other variables) to determine the minimum TAC of the system.

Example 11.3: Optimization of a Single-Stage RON

Consider the aqueous waste described in Example 11.1. ($Q_F = 26$ m^3/hr, $C_F = 0.035$ kmol/m^3 and $\pi_F = 1.57$ atm.). A 30 atm (PF) booster pump is used to pressurize the feed. Hollow fiber modules are to be configured in parallel (single stage) with the feed distributed equally among the units. The following properties are available for the HFRO modules:

$$\text{Permeability } (A) = 5.573 \times 10^{-8} \frac{m}{s \cdot atm}$$

$$\text{Salt flux constant } \left(\frac{D_{2M}}{K\delta} \right) = 1.82 \times 10^{-8} \text{ m/s}$$

$$\text{Viscosity } (\mu) = 10^{-3} \frac{kg}{m \cdot s}$$

$$\text{Outside radius of fibers } (r_o) = 42 \times 10^{-6} \text{ m}$$

$$\text{Inside radius of fibers } (r_i) = 21 \times 10^{-6} \text{ m}$$

$$\text{Fiber length } (L) = 0.750 \text{ m}$$

$$\text{Seal length } (L_S) = 0.075 \text{ m}$$

$$\text{Membrane area per module } (S_m) = 180 \text{ m}^2$$

$$\text{Pressure drop per module } = 1 \text{ atm.}$$

$$\text{Shell side pressure drop per module } = 0.4 \text{ atm.}$$

The flowrate per module is bounded by the following minimum and maximum flows:

$$3.0 \times 10^{-4} \leq q_F \left(\frac{m^3}{s \cdot \text{module}} \right) \leq 2.2 \times 10^{-3}. \tag{11.22}$$

The system cost can be evaluated via the following expressions:

Annualized fixed cost of modules including membrane replacement ($/yr)

$$= 2,300 \times \text{Number of modules} \tag{11.23}$$

Annualized fixed cost of pump ($/yr)

$$= 6.5[Q_F(P_F - 1.013 \times 10^5)]^{0.65}, \tag{11.24}$$

where Q_F is in m^3/s and P_F is in N/m^2.

Annualized fixed cost of turbine ($/yr) $= 18.4[Q_R(P_R - 1.013 \times 10^5)]^{0.43}$,

$$\tag{11.25}$$

where Q_R is in m³/s and P_R is in N/m²

Annual operating cost of pump ($/yr)

$$= \frac{Q_F(P_F - 1.013 \times 10^5)}{\eta_{Pump}} \times \frac{\$0.06/kWhr \times 8760\ hr/yr}{10^3\ W/kW}, \quad (11.26)$$

where Q_F is in m³/s, P_R is in N/m², and $\eta_{Pump} = 0.7$.

Annual operating cost of pretreatment (chemicals)

$$= \$0.03/m^3 \text{of feed} \quad (11.27)$$

Operating Value of Turbine ($/yr) $= Q_R(P_R - 1.013 \times 10^5)$

$$\times \eta_{Turbine} \frac{0.06}{10^3} \times 8760 \quad (11.28)$$

where Q_R is in m³/s and P_R is in N/m² and $\eta_{Turbine} = 0.7$.

TAC = Annualized fixed cost of modules + Annualized fixed cost of pump

+ Annualized fixed cost of turbine + Annual operating cost of pump

+ Annual operating cost of chemicals − Operating value of turbine

$$(11.29)$$

Design a minimum-TAC RON that can produce at least 10.7 m³/hr of permeate at a maximum allowable permeate composition of 0.0012 kmol/m³ (70 ppm).

Solution

Let us apply the above-mentioned procedure. We first select an iterative value of C_p. For instance, let us set $C_p = 0.0008$ kmol/m³. According to Eq. (11.17):

$$Q_R = Q_F - Q_P = 26 - 10.7 = 15.3\ \text{m}^3/\text{hr}.$$

Using the component material balance (Eq. 11.21):

$$C_R = \frac{Q_F C_F - Q_P C_P}{Q_R} = \frac{26.0 \times 0.0350 - 10.7 \times 0.0008}{15.3} = 0.0589\ kmol/m^3$$

Using Eq. (11.6), we have

$$C_S = \frac{0.0350 + 0.0589}{2} = 0.04695\ kmol/m^3.$$

The solute flux can be determined from Eq. (11.7)

$$N_{solute} = (1.82 \times 10^{-8})(0.04695) = 8.545 \times 10^{-10}\ \frac{kmol}{m^2 s}.$$

The water flux can be determined from Eq. (11.9a) as follows:

$$N_{water} = \frac{8.545 \times 10^{-10}}{0.0008} = 1.068 \times 10^{-6} \, m/s$$

Pressure difference across the membrane can be calculated using Eq. (11.5):

$$1.068 \times 10^{-6} = 5.573 \times 10^{-8} \left[\Delta P - \frac{1.57}{2} \left(1 + \frac{0.0589}{0.035} \right) \right] 0.6943,$$

i.e.,

$$\Delta P = 29.7 \, atm.$$

Therefore,

$$P_F = 29.7 + \left(\frac{0.4}{2} + 1.0 \right)$$

$$= 30.9 \, atm.$$

and from Eq. (11.9b), the volumetric flowrate of permeate per module:

$$q_p = S_m N_{water} = 180 \times 1.068 \times 10^{-6} = 1.92 \times 10^{-4} \frac{m^3}{module \, s}.$$

Hence, the number of modules can be calculated as follows:

$$n = \frac{Q_P}{q_P} = \frac{10.7}{3600 \times 1.92 \times 10^{-4}} = 15.5 \approx 16 \, modules.$$

We can now estimate the TAC of the system using Eqs. (11.23)–(11.29):

Annualized fixed cost of modules, $/yr = 2300 \times 16 = 36,800$ $/yr

Annualized fixed cost of pump, $/yr

$$= 6.5 \left[\frac{26}{3600} (30.9 \times 1.013 \times 10^5 - 1.013 \times 10^5) \right]^{0.65}$$

$$= 4304 \, \$/yr$$

Annualized fixed cost of turbine, $/yr

$$= 18.4 \left\{ \frac{15.3}{3600} [(30.9 \times 0.4) \times 1.013 \times 10^5 - 1.013 \times 10^5] \right\}^{0.43}$$

$$= 1070 \, \$/yr$$

Annual operating cost of pump, $/yr

$$= \frac{26(30.9 \times 1.013 \times 10^5 - 1.013 \times 10^5)}{3600 \times 0.7} \times \frac{0.06 \times 8760}{1000}$$

$$= 16,425 \, \$/yr$$

Annual operating cost of pretreatment chemicals

$$= 0.03 \times 8760 \times 26 = 6833 \text{ \$/yr}$$

Operating value of turbine, \$/yr

$$= \frac{15.3}{3600}[(30.9 - 0.4) \times 1.013 \times 10^5 - 1.013 \times 10^5]\frac{0.7 \times 0.06 \times 8760}{1000}$$

$$= 4672 \text{ \$/yr.}$$

Therefore,

$$\text{TAC} = 36800 + 4304 + 1070 + 16425 + 6833 - 4672 = 60760 \text{ \$/yr.}$$

Next, a new value of C_p is selected and the iterative scheme is continued. The results of the iterations are shown in Figs. 11.5 and 11.6 as functions of C_p and P_F, respectively.

As can be seen from the plots, the minimum TAC is about \$55,000/yr. The optimal permeate composition is about 0.0005 kmol/m^3 which corresponds to a feed pressure of 48 atm. It is interesting to note that the optimum value of C_p is significantly less than the required target composition (0.0012 kmol/m^3). In other words, more separation can be obtained for less cost! It is also worth mentioning that in some cases, environmental regulations may allow the bypass of a fraction of the feed and later mix it with the "over-separated permeate" to attain the required target composition. In such cases, lower costs than the ones shown in Figs. 11.5 and 11.6 can be accomplished.

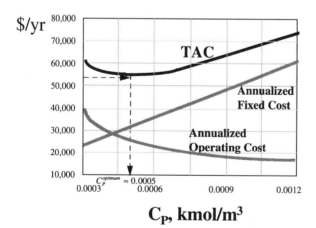

Figure 11.5 Cost versus permeate composition for Example 11.3.

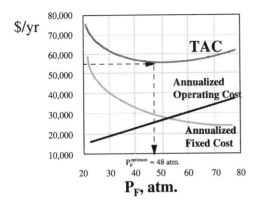

Figure 11.6 Cost versus feed pressure for Example 11.3.

Problems

11.1 An HFRO module with a membrane area of 300 m² is used to remove methyl ethyl ketone (MEK) from wastewater. The feed to the module has a flowrate of 9×10^{-4} m³/s, an MEK composition of 9500 ppm, and an osmotic pressure of 5.2 atm. The average pressure difference across the membrane is 50 atm. The permeate is collected at atmospheric pressure. The water recovery for the module is 86%, and the solute rejection is 98%. Evaluate the transport parameters $A\gamma$ and $(D_{2M}/K\delta)$.

11.2 Urea is removed from wastewater using a thin-film polysulfone reverse-osmosis membrane. When pure water was fed to the module under 68 atm., a permeate flux of 0.81 m³/m²·day was obtained. For a feed pressure of 68 atm and a feed composition of 1 wt/wt% of urea, it was found that the membrane provides a permeate flux of 0.56 m³/m²·day and a solute rejection of 85% (Matsuura, 1994). Evaluate the transport parameters $A\gamma$ and $(D_{2M}/K\delta)$ as well as the water recovery of the system.
Hint: For pure-water experiments, $A\gamma = N_{Water}/\Delta P$.

11.3 In the previous problem, it is desired to treat 12 m³/hr of wastewater containing 1 wt/wt% urea. The osmotic pressure of the feed is 7 atm. The power consumption of the system is estimated to be 5.0 kWh/m³ of permeate. Maintenance, pretreatment, and operating cost (excluding power) is \$0.2/m³ of permeate. The annualized fixed cost of the system may be evaluated through the following expression:

$$\text{Annualized fixed cost (\$/yr)} = 18,000 \text{ (Feed flowrate, m}^3\text{/hr)}^{0.7}.$$

Estimate the total annualized cost of the system and the cost per m³ of permeate.

11.4 In Example 11.1, the water recovery (permeate flowrate/feed flowrate) was 0.41. It is desired to increase water recovery by designing a multistage system. Two additional RO stages are to be used to process the reject from the first stage. Figure 11.7 illustrates

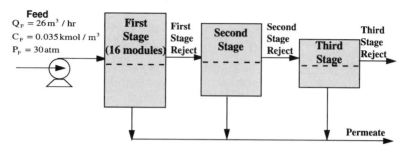

Figure 11.7 A three-stage tapered system for Problem 11.4.

the system configuration. The flowrate per module in the second and third stages is kept the same as in the first stage (4.5×10^{-4} (m^3/module · s)). As a result of permeation, the flowrate entering each stage decreases successively. Hence, the number of modules also decreases successively from one stage to the next, resulting in the tapered configuration shown in Fig. 11.7. Evaluate the total water recovery of the system and the average permeate composition.

Hint: Reject flowrate, composition, and pressure for the first stage become feed flowrate, composition, and pressure (respectively) for the second stage.

11.5 A wastewater stream has a flowrate of 22 m^3/hr and contains 26.0 ppm of monochlorophenol (MCP) and 3.0 ppm of trichlorophenol (TCP). Both phenolic compounds are toxic and must be reduced to acceptable levels prior to discharge. Design a cost-effective RON that satisfies the following specifications:

Minimum water recovery = 70%
Maximum composition of MCP in permeate = 9.5 ppm
Maximum composition of TCP in permeate = 1.2 ppm

The following data are available (El-Halwagi, 1992):

Permeability $(A) = 1.20 \times 10^{-8}$ kg/s · N
Solute flux constant for MCP = 2.43×10^{-4} kg/s · m^2
Solute flux constant for TCP = 2.78×10^{-4} kg/s · m^2
Outside radius of fibers = 42×10^{-6} m
Inside radius of fibers = 21×10^{-6} m
Fiber length = 0.750 m
Seal length = 0.075 m
Membrane area per module = 180 m^2
Pressure drop per module = 0.405×10^5 N/m^2
Permeate pressure = 1.013×10^5 N/m^2. The flowrate per module is bounded by the following minimum and maximum flows:

$$2.1 \times 10^{-4} \leq q_F \left(\frac{m^3}{s \cdot module} \right) \leq 4.6 \times 10^{-4}.$$

The system cost can be evaluated via the following expressions:

Annualized fixed cost of modules including membrane replacement ($/yr) $= 1450 \times$ Number of modules

Annualized fixed cost of pump ($/yr) $= 6.5[Q_F(P_F - 1.013 \times 10^5)]^{0.65}$, where Q_F is in m^3/s and P_F is in N/m^2.

Annualized fixed cost of turbine ($/yr) $= 18.4[Q_R(P_R - 1.013 \times 10^5)]^{0.43}$, where Q_R is in m^3/s and P_R is in N/m^2.

Annual operating cost of pump ($/yr)

$$= \frac{Q_F(P_F - 1.013 \times 10^5)}{\eta_{Pump}} \times \frac{\$0.06/kWhr \times 8760 \ hr/yr}{10^3 \ W/kW},$$

where Q_F is in m^3/s, P_R is in N/m^2, and $\eta_{Pump} = 0.7$.

Annual operating cost of pretreatment (chemicals) $= \$0.03/m^3$ of feed.

11.6 Consider the textile process shown in Fig. 11.8 (Muralikrishnan *et al.*, 1996). Prior to spinning the fibers, a sizing agent is deposited onto the fibers to enhance their processability during spinning. The sizing agent in this problem is polyvinyl alcohol (PVA). PVA is added as a 5.5% (by wt) solution in water. During spinning operation part of the water that is added along with PVA is lost to the surroundings. After spinning, the sizing agent has to be removed from the fibers by washing the fibers with hot water in a desizing unit. The fibers are washed again in a washing unit to remove any residual PVA. Fibers leaving the washing unit pass through other finishing operations. Figure 11.8 shows the pertinent material balance data. The wastewater stream of the plant consists of water from desizing and washing units. This wastewater stream is fed to the biotreatment unit.

Because of the high value of PVA, the plant is interested in reducing PVA losses through mass integration. Using the tools described in Chapters Four, Seven, and Eleven, devise cost-effective strategies for maximizing the value of recovered PVA—recovery cost. The devised strategies may include segregation, mixing, recycling and interception using HFRO. The following information may be used in design:

Recycle Constraints

The following technical constraints should be observed in any proposed solution:

(a) Desizing:

- $7.0 \leq$ Flow rate of desize feed (kg/s) ≤ 7.7
- $0.00 \leq$ PVA concentration (wt%) ≤ 0.15

(b) Washing:

- $8.5 \leq$ Flow rate of wash feed (kg/s) ≤ 9.0
- $0.00 \leq$ PVA concentration (wt%) ≤ 0.08

(c) Sizing:

- $1.5 \leq$ Flow rate of sizing feed (kg/s) ≤ 1.6
- $5.6 \leq$ Concentration of PVA (wt%) ≤ 5.7

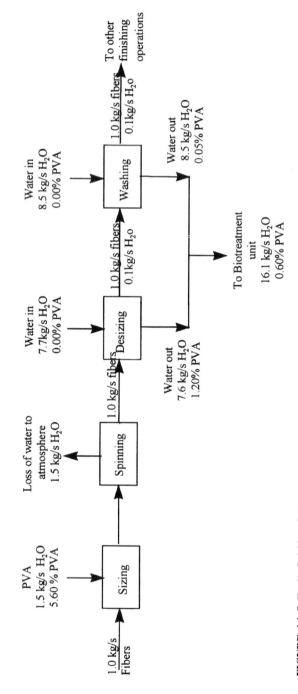

FIGURE 11.8 Textile finishing plant.

Separation of PVA from Water

HFRO modules may be used for separating PVA from water. The data required for designing HFRO are given below.

(a) Properties of HFRO modules:

Permeability $(A\gamma) = 5.573 \times 10^{-8}$ m/(s · atm)
Solute constant $(D_{2M}/k\delta) = 3.047 \times 10^{-8}$ m/s
Outside radius of fibers $(r_o) = 42 \times 10^{-6}$ m
Inside radius of fibers $(r_i) = 21 \times 10^{-6}$ m
Fiber length (L) = 0.750 m
Seal length $(L_s) = 0.075$ m
Membrane area per module $(S_m) = 180$ m^2
Pressure drop per module $(P_d) = 0.4$ atm
Maximum allowable flow rate per module $(q_{f,max}) = 2.2 \times 10^{-3}$ m^3/(module · s)
Minimum allowable flow rate per module $(q_{f,min}) = 3.00 \times 10^{-4}$ m^3/(module · s)
Maximum allowable pressure $= 44.0 \times 10^5$ N/m^2
Osmatic pressure (atm) = 5.66 [concentration (in wt%)]
Molecular weight of PVA = 100000 kg/kg mol

(b) Cost Data:

Annualized fixed cost of modules including membrane replacement ($/yr) = 5000 × Number of modules
Annualized fixed cost of pump ($/yr) $= 6.5[Q_F(P_F - 1.013 \times 10^5)]^{0.65}$ where Q_F is in m^3/s and P_F is in N/m^2
Annualized fixed cost of turbine ($/yr) $= 18.4[Q_R(P_R - 1.013 \times 10^5)]^{0.43}$ where Q_R is in m^3/s and P_R is in N/m^2
Annualized fixed cost of piping $= 50/m \cdot yr$
Annual operating cost of pump

$$(\$/yr) = \frac{Q_F(P_F - 1.013 \times 10^5)}{\eta_{pump}} \times \frac{\$0.06/kWhr \times 8760\ hr/yr}{10^3\ W/kW}$$

where Q_F is in m^3/s and P_F is in N/m^2
Annual operating cost of pretreatment (chemicals) = \$0.03/m^3 of feed
Operating value of Turbine ($/yr) $= Q_R(P_R - 1.013 \times 10^5)\eta_{Turbine} \times 0.06/10^3 \times 8760$
where Q_R is in m^3/s and P_R is in N/m^2 and $\eta_{turbine} = 0.7$
PVA cost = \$2.2/kg
Cost of wash water including heating = \$9/m^3
Wastewater treatment cost = \$2/m^3

Symbols

A	Permeability (m/s · atm)
C_F	Concentration of solute in feed stream (kmol/m^3)
C_P	Concentration of solute in permeate stream (kmol/m^3)
C_R	Concentration of solute in reject stream (kmol/m^3)

C_S	Average solute concentration in shell side ($kmol/m^3$)
L	Fiber length (m)
L_S	Seal length (m)
n	Number of modules
N_{solute}	Solute flux through the membrane ($kmol/m^3$)
N_{water}	Water flux through the membrane ($kmol/m^3$)
P_F	Pressure of feed stream entering a module (atm)
P_P	Pressure of permeate stream leaving a module (atm)
P_R	Pressure of reject stream leaving a module (atm)
ΔP	Pressure difference across the membrane (atm)
q_F	Feed flow rate per module ($m^3/s \cdot$ module)
q_P	Permeate flow rate module ($m^3/s \cdot$ module)
q_R	Reject flow rate per module ($m^3/s \cdot$ module)
Q_F	Feed flow rate (m^3/s)
Q_P	Flow rate of permeate (m^3/s)
Q_R	Flow rate of reject (m^3/s)
r_i	Inside radius of fibers (m)
r_o	Outside radius of fibers (m)
S_m	Surface area of hollow fibers per module (m^2)

Greek letters

γ	a constant defined by Eq. (11.4)
η	a constant defined by Eq. (11.4)
θ	a constant defined by Eq. (11.4)
μ	Viscosity ($kg/m \cdot sec$)
π_F	osmatic pressure of feed (atm)
μ_{Pump}	Efficiency of pump
$\eta_{Turbine}$	Efficiency of turbine

Subscripts

F	Feed
P	Permeate
R	Reject

Superscripts

min	minimum
max	maximum

References

Baker, R. W., Koros, W. J., Cussler, E. L., Riley, R. L., Eykamp, W., and Strathmann, H. (1991). "Membrane Separation Systems: Recent Development and Future Directions," Noyes Data Corp, Park Ridge, NJ.

Belfort, G. (1984) "Synthetic Membrane Processes," Academic Press, Orlando, FL.

Bitter, J. G. A. (1991) "Transport Mechanisms in Membrane Separation Processes," Plenum Press, New York.

Comb, L. F. (1994) Going forward with reverse osmosis. *Chem. Eng.*, July, pp. 90–92.

Crabtree, E. W., Dunn, R. F., and El-Halwagi, M. M. (1997). Synthesis of hybrid gas permeation membrane/condensation systems for pollution prevention. *J. Air Waste Manage.* Association (in press).

Crespo, J., and Böddeker, K. W. (1994). "Membrane Processes in Separation and Purification," Kluwer Academic Publisher, Dordrecht, The Netherlands.

Dandavati, M. S., Doshi, M. R., and Gill, W. N. (1975). Hollow fiber reverse osmosis: Experiments and analysis of radial flow systems. *Chem. Eng. Sci.*, 30, 877–886.

El-Halwagi, M. M. (1992). Synthesis of reverse osmosis networks for waste reduction, *AIChE J.*, 38(8), 1185–1198.

El-Halwagi, M. M. (1993). Optimal design of membrane hybrid systems for waste reduction. *Sep. Sci. Technol.*, 28(1–3), 283–307.

El-Halwagi, A. M., and El-Halwagi, M. M. (1992). Waste minimization via computer-aided chemical process synthesis—A new design philosophy. *TESCE*, 18(2), 155–187.

El-Halwagi, A. M., Manousiouthakis, V., and El-Halwagi, M. M. (1996). Analysis and simulation of hollow fiber reverse osmosis modules. *Sep. Sci. Tech.*, 31(18), 2505–2529.

Evangelista, F. (1986). Improved graphical analytical method for the design of reverse osmosis desalination plants. *Ind. Eng. Chem. Process Des. Dev.*, 25(2), 366–375.

Gupta, S. K. (1987). Design and analysis of a radial-flow hollow-fiber reverse-osmosis system. *Ind. Eng. Chem. Res.* 26, 2319–2323.

Hermans, J. J. (1978). Physical aspects governing the design of hollow fiber modules. *Desalination*, 26, 45–62.

Kabadi, V. N., Doshi, M. R., and Gill, W. G. (1979). Radial flow hollow fiber reverse osmosis: Experiments and theory. *Chem. Eng. Commun.* 3, 339–365.

Matsuura, T. (1994). "Synthetic Membranes and Membrane Separation Processes." CRC Press, Boca Raton, FL.

Mulder, M. (1991). "Basic Principles of Membrane Technology." Kluwer Academic Publishers, Dordrecht, The Netherlands.

Muralikrishnan, G., Crabtree, E., and El-Halwagi, M. M. (1996). "Design of Membrane Hybrid Systems for Pollution Prevention." AIChE Spring Meeting, New Orleans, LA.

Ohya, H., Nakajima, H., Takagi, K., Kagawa, S., and Negishi, Y. (1977). An analysis of reverse osmotic characteristics of B-9 hollow module. *Desalination.* 21, 257–274.

Orofino, T. A. (1977). Technology of hollow fiber reverse osmosis systems." In "Reverse Osmosis and Synthetic Membranes" (S. Sourirajan, eds.), pp. 313–341 National Research Council, Ottawa, Canada.

Parekh, B. S. (1988). "Reverse Osmosis Technology—Application for High-Purity Water Production." Dekker, New York.

Rautenbach, R., and Albrecht, R. (1989). "Membrane Processes." Wiley, New York.

Rorech, G. J. and Bond, S. G. (1993). "Reverse Osmosis: A Cost Effective Versatile Water Purification Tool," I&EC, pp. 35–37.

Sourirajan, S., and Matsuura, T. (1985). "Reverse Osmosis/Ultrafiltration Process Principles." National Research Council, Ottawa, Canada.

Soltanieh, M., and Gill, W. N. (1982). Analysis and design of hollow fiber reverse osmosis systems. *Chem. Eng. Commun.*, **18**, 311–330.

Soltanieh, M., and Gill, W. N. (1984). An experimental study of the complete mixing model for radial flow hollow fiber reverse osmosis systems. *Desalination*, **49**, 57–88.

Weber, W. J., Jr. (1972). "Physicochemical Processes for Water Quality Control," p. 312. Wiley Interscience, New York.

Zhu, M., El-Halwagi, M. M., and Al-Ahmad, M. (1997). Optimal design and scheduling of flexible reverse osmosis networks. *J. Membr. Sci.*, (in press).

CHAPTER TWELVE

Environmentally Benign Chemistry and Species

An important element of pollution prevention is the selection of environmentally benign chemical reactions, raw materials, solvents, and products. Over the past few years, significant progress has been made in this area. This chapter provides a brief overview of the recent advances in synthesizing "green" reactions and species. For more detailed discussion, the reader is referred to Anastas and Williamson (1996), Anastas and Farris (1994), and Chase (1995).

12.1 Synthesis of Environmentally Acceptable Reactions

For a given desired product, there are typically numerous reaction alternatives that should be identified and screened. The identification of these reactions is not a straightforward task. The problem is further compounded when the search is limited to environmentally benign cost-effective chemistry. In this context, the following questions should be answered:

- What raw materials should be employed?
- What is the yield of the desired product?
- What is the distribution of by-products? Are all by-products environmentally acceptable?
- How can undesirable species be removed or minimized?
- What are the operating conditions that ensure thermodynamic feasibility of these reactions?
- Is a catalyst needed to achieve desired kinetics? What catalyst should be employed?
- How should the reaction system by sized and configured?

In order to address these challenges, a hierarchical approach may be adopted. This approach focuses on the "big picture" first, then adds details to promising solutions. Therefore, preliminary screening ought to be conducted first to identify overall reaction alternatives that meet process requirements in terms of desired product, cost effectiveness, environmental acceptability, and thermodynamic feasibility. At this stage, minimum details are to be invoked. The problem of synthesizing environmentally acceptable reactions "EARs" has been introduced by Crabtree and El-Halwagi (1994) and can be stated as follows:

> Given a reactor of known size and functionality, and a desired product along with its flowrate, synthesize an overall chemical reaction that features maximum economic potential while complying with all environmental and thermodynamic constraints.

In order to generate a candidate EAR, one should consider potential raw materials and by-products, satisfaction of stoichiometric conditions, assurance of thermodynamic feasibility, and fulfillment of environmental requirements. These issues can be addressed by employing an optimization formulation to identify an overall reaction that yields the desired product at maximum economic potential while satisfying stoichiometric, thermodynamic, and environmental constraints. For a more detailed description of this optimization program, the reader is referred to Crabtree and El-Halwagi (1994).

In order to shed more light on synthesizing EARs, it is instructive to discuss the following case study on producing 1-naphthyl-N-methyl carbamate (or carbaryl). This product was produced in Bhopal (India) by reacting α-naphthol and methyl isocyanate "MIC." In 1984, approximately 50 tons of MIC underwent a chemical reaction and leaked into the atmosphere forming a toxic cloud which killed over 2,000 people and injured more than 250,000. It is possible that this tragic accident could have been averted if carbaryl had been produced from a more environmentally acceptable reaction. Let us consider the production of 3 tons/hr of carbaryl in a 30 m^3 CSTR. By applying the EARs synthesis procedure of Crabtree and El-Halwagi (1994), the following optimal EAR is identified:

$$\underset{\text{Methyl Formamide}}{H-\underset{\underset{O}{\|}}{C}-\overset{\overset{H}{|}}{N}-CH_3} \; + \; \underset{\alpha-\text{Naphthol}}{\text{(naphthol)}\,O-H} \; = \; \underset{\text{Carbaryl}}{\text{(naphthyl)}\,O-\underset{\underset{O}{\|}}{C}-\overset{\overset{H}{|}}{N}-CH_3} \; + \; \underset{\text{Hydrogen}}{H_2}$$

(12.1)

This reaction has an economic potential of $40.05 million per year. An advantage of the optimization-based approach to synthesizing EARs is its ability to generate next-to-optimal solutions. This can be achieved by adding constraints to the optimization program in order to guarantee that the previous solution is not obtained again. Hence, by excluding the optimal EAR and solving the optimization program, we get the following reactions:

$$0.5 \text{ Oxygen} + \text{Methyl Formamide} + \alpha-\text{Naphthol} = \text{Carbaryl} + \text{Water} \quad (12.2)$$

which yields an economic potential of \$39.99 million per year, and

$$\text{Methyl Amine} + \alpha\text{-Naphthol Chloroformate} = \text{Carbaryl} + \text{Hydrogen Chloride.} \quad (12.3)$$

This reaction, while thermodynamically feasible and environmentally acceptable, shows no sign of making a profit. It corresponds to a gross revenue of \$-44,300 per hour; a major loss. Hence, no additional effort should be invested in investigating kinetics, catalysis, or reactor design for this reaction. On the other hand, reactions (12.1) and (12.2) are promising candidates that should be experimentally investigated to validate their feasibility and fine tune their optimal conditions. Once these overall reactions are chosen, it is important to undertake studies on reaction path synthesis to identify the reaction mechanism and the elemental steps involved (e.g., Chase, 1995; Knight and McRae, 1993; Douglas, 1992; Mavrovouniotis *et al.*, 1990, 1992; Fornari *et al.*, 1989; Rotstein *et al.*, 1982; Govind and Powers, 1981; Rudd and May, 1976; Rudd, 1976; Rudd *et al.* 1973). Next, catalysis and kinetics should be investigated to manipulate the reaction rates (e.g., Dartt and Davis, 1994; Haggin, 1994; Sinfelt, 1987; Boudart and Djega-Mariadassou, 1984; El-Halwagi, 1971). Finally, reactor design and optimization is undertaken and linked with separation systems (e.g., Muralikrishnan *et al.*, 1996; Sund and Lien, 1996; Friedler *et al.*, 1994; Balakrishna and Biegler, 1992; Glasser *et al.*, 1992; Kokossis and Floudas, 1991; Hildebrandt *et al.*, 1990; El-Halwagi, 1990; El-Halwagi and El-Rifai, 1988; Glasser *et al.*, 1987; Fogler, 1986; Levenspiel, 1972; Horn, 1965; Fan *et al.*, 1965). This hierarchical approach ensures that only relevant details are examined at each stage of the analysis and that only promising reaction routes are investigated.

12.2 Synthesis of Environmentally Benign Species

In many cases, it is possible to replace environmentally hazardous chemicals with more benign species without compromising the technical and economic performance of the process. Examples include alternative solvents, polymers, and refrigerants. Group contribution methods have been commonly used in predicting physical and chemical properties of synthesized materials. Two main frameworks have been employed to synthesize alternative materials: knowledge base and computer-aided optimization. Knowledge-based approaches depend on understanding the criteria of the materials to be replaced along with general rules and algorithms that link properties with structure. Examples of this approach can be found in literature (e.g., Joback, 1994; Joback and Stephanopoulos, 1990; Constantinou *et al.*, 1994) proposing an algorithm to synthesize molecules subject to a set of property constraints. Computer-aided optimization approaches are based on formulating the molecular design problem as an optimization program which seeks to maximize a performance function or minimize deviation from desired properties subject to various constraints including structural feasibility, property structure correlations, and environmental criteria. Examples of this approach include synthesis of solvents (e.g., Brignole *et al.*, 1986; Odele and Macchietto, 1990; Naser and Fournier, 1991; Dunn *et al.*, 1997), polymers (e.g.,Vaidyanathan and El-Halwagi, 1994 and 1996; Venkatasubramanian *et al.*, 1994), and refrigerants (e.g., Achenie and Duvedi, 1996).

Process synthesis techniques can be integrated with product synthesis tools. As an illustration, let us revisit the problem of synthesizing mass-exchange networks (MENs) described in Chapter Three. A key condition for applying the mass-pinch techniques of Chapter Three is that the designer has a set of candidate mass-separating agents (MSAs). Therefore, it may be necessary to solve a product synthesis problem to generate the candidate MSAs needed for the mass-pinch approach. This problem of simultaneously synthesizing MSAs and MENs has been introduced by Hamad *et al.* (1996) and can be stated as follows: "Given a set of chemical functional groups, it is desired to synthesize a set of MSAs to selectively transfer certain targeted species (e.g., pollutants) from a set of rich streams at minimum cost. These MSAs must meet technical, environmental, and safety constraints. Figure 12.1 is a schematic representation of the problem statement.

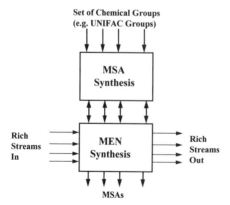

Figure 12.1 Schematic representation of simultaneous synthesis of MSAs and MEN (Hamad *et al.*, 1996).

First, candidate chemical groups are chosen as building blocks for the solvents. A common choice is the UNIFAC functional groups. These groups are, in general, classified into three categories: extenders, branches, and terminators. Special rules are available for linking these groups in a structurally feasible manner (e.g., Gani *et al.*, 1990; Vaidyanathan *et al.*, 1994, 1996). Next, property structure correlations such as group-contribution methods are selected. The fundamental assumption of most group-contribution methods is that the contribution toward any property made by one group is independent of that made by another group. Mass-exchange equilibrium may be predicted using thermodynamic models such as the UNIFAC method and the regular solution theory. Finally, MSA solvent constraints are integrated with MEN synthesis formulation in a mixed-integer nonlinear program which minimizes the cost of the MSAs and the MEN subject to MSA-synthesis constraints, MEN synthesis constraints, technical requirements as well as safety and environmental restrictions.

As an example, let us consider the recovery of butane from a methane-rich stream in a benzene production facility (Hamad *et al.*, 1996). It is required to synthesize a minimum cost system of MSAs and MEN. The following building blocks of chemical groups were chosen: CH_3, CH_2, CH, C, OH, $COOH$, CHO, $CH_2=CH$, $CH=CH$, $CH=C$, CH_2Cl, $CHCl$, CH_2NH_2, CH_3NH, ACH, AC, $ACCH_3$, $ACOH$, and $ACCl$ (where A denotes any aromatic group).

The solution to this case study is to use cycle oil (a process MSA) followed by a synthesized external MSA (heptane) as shown in Fig. 12.2.

A key advantage of optimization-based approaches is their ability to generate a list of candidate solutions. Once an optimal solution is determined, the chosen MSA can be removed from the search. Resolving the optimization problem, we then get another candidate MSA. By successively eliminating identified solutions, one can generate a large list of potential MSAs. It is worth noting that this procedure is intended to give a conceptual framework for generating promising solutions. Once these solutions are determined, more detailed analysis is needed including

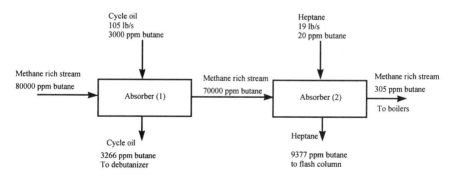

Figure 12.2 Solution of the butane-recovery example (Hamad *et al.*, 1996).

thorough thermodynamic modeling and experimental verification. The advantage of this conceptual framework is to generate candidates that are not intuitively apparent and to focus the designers effort on a small set of promising solutions. This is quite consistent with the overall philosophy of this book; breadth first, depth later.

References

Achenie, L. E. K., and Duvedi, A. P. (1996). Designing environmentally safe refrigerants using mathematical programming. *Chem. Eng. Sci.*, **51**(15), 3727–3739.

Anastas, P. T., and Farris, C. A. (eds.) (1994). "Benign by design: Alternative synthetic design for pollution prevention," ACS Symp. Ser., vol. **577**, ACS Pub., Washington, DC.

Anastas, P. T., and Williamson, T. C. (eds.) (1996). "Green chemistry: Designing chemistry for the environment," ACS Symp. Ser., vol. **626**, ACS Pub., Washington, DC.

Balakrishna, S., and Biegler, L. T. (1992). "Targeting strategies for the synthesis and energy integration of nonisothermal reactor networks," *Ind. Eng. Chem. Res.*, **31**, 2152–2164.

Boudart, M., and Djega-Mariadassou, G. (1984). "Kinetics of Heterogeneous Catalytic Reactions," Princeton University Press, Princeton, NJ.

Brignole, E. A., Bottini, S., and Gani, R. (1986). A strategy for the design and selection of solvents for separation processes. *Fluid Phase Equilibria*, **29**, 125–132.

Chase, V. (1995). Green chemistry: The middle way to a cleaner environment. *R & D Magazine*, 25–26.

Constantinou, L., Jacksland, C., Bagherpour, K., Gani, R., and Bogle, L. (1994). Application of group contribution approach to tackle environmentally-related problems. *AIChE Symp. Ser.*, **90**(303), 105–116.

Crabtree, E. W., and El-Halwagi, M. M. (1994). Synthesis of environmentally-acceptable reactions. *AIChE Symp. Ser.*, **90**(303), 117–127.

Dartt, C. B., and Davis, M. E. (1994). Catalysis for environmentally benign protection. *Ind. Eng. Chem. Res.*, **33**, 2887–2899.

Denbigh, K. G., and Turner, J. C. R. (1984). "*Chemical Reactor Theory,*" Cambridge University Press: Cambridge.

Douglas, J. M. (1992). Process synthesis for waste minimization. *Ind. Eng. Chem. Res.*, **31**, 238–243.

Dunn, R. F., Dobson, A. M., and El-Halwagi, M. M. (1997). Optimal design of environmentally acceptable solvent blends for coatings. *Adv. Env. Res.*, **1**(2), in press.

Dunn, R. F., El-Halwagi, M. M., Lakin, J., and Serageldin, M. (1995). Selection of organic solvent blends for environmental compliance in the coating industries. Proceedings of the First International Plant Operations and Design Conference, E. D. Griffith, H. Kahn and M. C. Cousins (eds.). Vol. III, pp. 83–107, AIChE, New York.

El-Halwagi, M. M. (1971). An Engineering Concept of Reaction Rate. *Chem. Eng.*, 75–78.

El-Halwagi, M. M., (1990). Optimization of bubble column slurry reactors via natural delayed feed addition,"*Chem. Eng. Comm.*, **92**,103–119.

El-Halwagi, M. M., and El-Rifai, M. A. (1988). Mathematical modeling of fluidized-bed reactors—I. The multistage three phase model," *Chem. Eng. Sci.*, **43**(9), 2477–2486.

Fan, L. T., Erickson, L. E., Sucher, R. W., and Mathad, G. S. (1965). Optimal design of a sequence of continuous-flow stirred-tank reactors with product recycle. *Ind. Eng. Chem.*, **4**, 432–440.

Fogler, S. H. (1992). "Elements of Chemical Reaction Engineering," Prentice Hall, Englewood Cliffs, N.J.

Fornari, T., Rotstein, E., and Stephanopoulos, G. (1989). "Studies on the Synthesis of Chemical Reaction Paths — II. Reaction Schemes with Two Degrees of Freedom," *Chem. Eng. Sci.*, **44**(7), 1569–1579.

Friedler, F., Varga, J. B., and Fan, L. T. (1994). Algorithmic approach to the integration of total flowsheet synthesis and waste minimization. *AIChE Symp. Ser.*, **90**(303), 86–97.

Glasser, B., Hildebrandt, D., and Glasser, D. (1992). Optimal mixing for exothermic reversible reactions. *Ind. Eng. Chem. Res.*, **31**, 1541–1549.

Glasser, D., Hildebrandt, D., and Crowe, C. (1987). A geometric approach to steady flow reactors: The attainable region and optimization in concentration space. *Ind. Eng. Chem. Res.*, **26**, 1803–1810.

Govind, R., and Powers, G. J. (1981). Studies in reaction path synthesis. *AIChE J.*, **27**(3), 429–442.

Haggin, J. (1994). Catalysis gains widening role in environmental protection. *Chem. Eng. News*, **72**, 22–30.

Hamad, A. A., Gupta, R. B., Dunn, R. F., and El-Halwagi, M. M. (1996). Simultaneous synthesis of separating agents and separation networks for removal of VOCs from air. *AIChE Annu. Meet.*, Chicago.

Hildebrandt, D., Glasser, D., and Crowe, C. (1990). Geometry of the attainable region generated by reaction and mixing: With and without constraints. *Ind. Eng. Chem. Res.* **29**, 49–58.

Hopper, J. R., Yaws, C. L., Ho, T. C., and Vichailak, M. (1993). Waste minimization by process modification. *Waste Management*, **1**(13), 3–14.

Horn, F. (1965). Attainable and non-attainable regions in chemical reaction technique, in "*Third European Symposium on Chemical Reaction Engineering*," Pergammon Press: London, 293–302.

Joback, K. G. (1994). "Solvent Substitution for Pollution Prevention," *AIChE Symp. Ser.*, Vol. 90, Number 303, 98–104

Joback, K. G., and Stephanopoulos, G. (1990). Designing molecules possessing desired physical property values. In "Foundations of Computer Aided Process Design 'FOCAPD' III" (J. J. Siirola, I. Grossmann, and G. Stephanopoulos, eds.), pp. 363–387. CACHE/Elsevier, New York.

Kokossis, A. C., and Floudas, C. A. (1994). Optimization of complex reactor networks: Nonisothermal operation. *Chem. Eng. Sci.*, **49**, 1037–1051.

Levenspiel, O. (1972). "Chemical Reaction Engineering," Second Edition, John Wiley and Sons, New York.

Macchietto, S., Odele, O., and Omatsone, O. (1990). Design of optimal solvents for liquid–liquid extraction and gas absorption processes, *Trans Inst. Chem. Eng.*, **68**(A), 429–433.

Mavrovouniotis, M. L., and Stephanopoulos, G. (1990). Computer-aided synthesis of biochemical pathways. *Biotechnology and Bioengineering*, **36**, 1119–1132.

Mavrovouniotis, M. L., and Stephanopoulos, G. (1992). Synthesis of reaction mechanisms consisting of reversible and irreversible steps. *Ind. Eng. Chem. Res.*, **31**, 1625–1637.

Muralikrishnan, G., Ramdoss, P. and El-Halwagi, M. M. (1996). Integrated design of reaction and separation systems for waste minimization. *AIChE Annu. Meet.*, Chicago.

Naser, S. F., and Fournier, R. L. (1991). A system for the design of an optimum liquid-liquid extractant molecule. *Comp. Chem. Eng.*, **15**(6), 397–414.

Odele, O., and Macchietto, S. (1993). Computer aided molecular design: A novel method for optimal solvent selection. *Fluid Phase Equilibria*, **82**, 47–54.

Rotstein, E., Resasco, D., and Stephanopoulos, G. (1982). Studies on the synthesis of chemical reaction paths—I. Reaction characteristics in the (ΔG, T) space and a primitive synthesis procedure. *Chem. Eng. Sci.*, **37**(9), 1337–1352.

Rudd, D. F. (1976). Accessible designs in solvay cluster synthesis. *Chem. Eng. Sci.*, **31**, 701–703.

Rudd, D. F., and May, D. (1976). Development of solvay clusters of chemical reactions. *Chem. Eng. Sci.*, **31**, 59–69.

Rudd, D. F., Powers, G. J., and Siirola, J. J. (1973). "Process Synthesis," Prentice Hall, Inc., N.J.

Sinfelt, J. H. (1987). Catalyst design: Selected topics and examples. In "Recent developments in chemical process and plant design," Y. A. Liu, H. A. McGee and W. R. Epperly (eds.), John Wiley and Sons, New York, pp. 1–41.

Smith, W. R., and Missen, R. W. (1982). "Chemical Reaction Equilibrium Analysis," John Wiley and Sons, New York.

Sund, E. B., and Lien, K. (1996). An optimal control formulation of the reaction-mixing problem. *AIChE Annu. Meet.*, Chicago.

Vaidyanathan, R., and El-Halwagi, M. M. (1994). Computer-aided design of high performance polymers. *J. of Elastomers and Plastics*, **26**, 277–293.

Vaidyanathan, R., and El-Halwagi, M. M. (1996). Computer-Aided synthesis of polymers and blends with target properties. *Ind. Eng. Chem. Res.*, **35**, 627–634.

Venkatasubramanian, V., Chan, K., and Cauthers, J. M. (1994). Computer-aided molecular design using genetic algorithms. *Comp. Chem. Eng.*, **18**, 833–844.

Appendix I: Useful Relationships for Compositions

Pollutant concentration may be reported in various ways. It is very important to be able to relate the different composition units.

I.1 Mass versus Molar Compositions

Mass can be related to moles via molecular weight, i.e.,

$$n_i = \frac{m_i}{M_i},$$

(I.1)

where m_i is the mass of species i, M_i is the molecular weight of species i, and n_i is the number of moles of species i.

Therefore, molar ratios of two species, i and j, can be related to their mass ratios as follows:

$$\frac{n_i}{n_j} = \left(\frac{M_j}{M_i} \right) \left(\frac{m_i}{m_j} \right)$$

(I.2)

and mole fractions can be related to mass fractions as follows:

$$y_i = \left(\frac{\bar{M}}{M_i} \right) x_i,$$

(I.3)

where y_i is the mole fraction of species i, $\bar{M}$ is the average molecular weight of mixture $(\sum_{all\ species} y_i M_i)$, and x_i is the mass fraction of species i.

297

I.2 Gas Composition versus Partial Pressure

To relate partial pressures to compositions, one may use the ideal gas law:

$$p_i V = n_i RT \qquad \text{(I.4a)}$$

$$= \frac{m_i}{M_i} RT, \qquad \text{(I.4b)}$$

where V is volume and p_i is partial pressure of species i which can be related to total pressure of the gaseous mixture, P, via Dalton's law:

$$y_i = \frac{P_i}{P}. \qquad \text{(I.5)}$$

The term R is the gas constant which has the following values in different units:

8,314.3 m^3 Pa/kgmol K
8.3143 m^3 kPa/kgmol K
8.3143 J/gmol K
0.082057 m^3 atm/kg mol K
0.082057 lit atm/g mol K
82.057 cm^3 atm/g mol K
1.9872 cal/g mol K
1.9872 BTU/lbmol $^{\circ}R$
10.731 psia ft^3/lbmol $^{\circ}R$
21.9 ft^3 (inch Hg)/lb mol $^{\circ}R$
0.7302 atm ft^3/lb mol $^{\circ}R$
1,545.3 ft lb$_f$/lb mol $^{\circ}R$

Equation (I.4b) can be rearranged to give mass density of an ideal gas, ρ, as

$$\rho_i = \frac{M_i}{V_i} = \frac{p_i M_i}{RT} \qquad \text{(I.6)}$$

For instance, density of air at 1 atm and 298 K is

$$\rho_{\text{air}} = \frac{1 \times 29}{0.082057 \times 298} \approx 1.2 \text{ kg/m}^3.$$

I.3 Parts per Million

In describing concentrations of pollutants in gaseous and aqueous wastes, a commonly confusing issue is the definition of parts per million, ppm. Sometimes, it is specified that these are compositions based on weight, moles, or volume (ppmw, ppmm or ppmv, respectively). However, in most cases, it is reported as ppm. Depending on the phase of the stream, ppm implies the basis of weight for liquids and mole or volume for gases as explained in the following sections.

For Gases

$$\text{ppm of species i} = \frac{\text{mole of gaseous species i}}{\text{moles of gaseous mixture}} \times 10^6 \qquad \text{(I.7a)}$$

$$= \text{mole fraction of species i} \times 10^6, \qquad \text{(I.7b)}$$

where the mole fraction of species i in a gaseous mixture may be evaluated via Eq. (I.5).

Hence, one can combine Eqs. (I.4)–(I.7) to develop the following expression for converting from ppm in the gas phase to mass concentration of species i:

$$c_i = \frac{m_i}{V} = \frac{(\text{ppm of species i}) \, PM_i}{RT} \times 10^6. \qquad \text{(I.8)}$$

For instance, 1 ppm of ozone (O_3, molecular weight $= 48$) in air at 1 atm and 298 K can be converted into kg/m^3 as follows:

$$\text{mass concentration (kg/m}^3) = \frac{1 \times 1 \times 48 \times 10^{-6}}{0.082057 \times 298}$$

$$= 1.96 \times 10^{-6} \text{ kg/m}^3.$$

For Liquids

$$\text{ppm} = \frac{\text{mass of pollutant}}{\text{mass of liquid mixture}} \times 10^6 \qquad \text{(I.9a)}$$

$$= \text{mass fraction of pollutant in liquid} \times 10^6. \qquad \text{(I.9b)}$$

Of particular importance is the case of wastewater with dilute pollutants at ambient temperature. In this case, one can use the following conversion from ppm to mass concentration:

$$1 \text{ ppm} = 1 \text{ g/m}^3 \qquad \text{(I.10a)}$$

$$= 1 \text{mg/lit.} \qquad \text{(I.10b)}$$

Appendix II: Conversion Factors

Length

1 in.	= 2.540 cm
1 ft	= 0.3048 m
1 yd	= 0.9144 m
1 mile	= 1.6093 km
	= 5280 ft
1 m	= 3.2808 ft
	= 39.37 in.
1 km	= 0.6214 mile
1 micron	= 10^{-6} m
	= 10^{-4} cm
	= 10^{-3} mm
	= 1 μm (micrometer)
1 Å (angstrom)	= 10^{-10} m
	= 0.1 nm

Volume

1 L (Liter)	= 1000 cm^3
1 m^3	= 1000 L
	= 264.17 U.S. gal
1 U.S. gal	= 4 qt
	= 3.7854 L
1 ft^3	= 28.317 L
	= 0.028317 m^3
	= 7.481 U.S. gal
1 in.^{3}	= 16.387 cm^3
1 bbl (barrel)	= 5.6146 ft^3
	= 42 U.S. gal
	= 0.15899 m^3

Mass

1 lb$_m$	= 453.59 gm
	= 0.45359 kg
	= 16 oz
	= 7000 grains
1 kg	= 1000 gm
	= 2.2046 lb$_m$
1 ton (metric)	= 1000 kg
1 ton (short)	= 2000 lb$_m$
1 ton (long)	= 2240 lb$_m$
	= 1016 kg

Density

1 lb$_m$/ft^3	= 16.018 kg/m^3
1 kg/m^3	= 0.062428 lb$_m$/ft^3
1 gm/cm^3	= 62.427961 lb$_m$/ft^3

Pressure

1 atm	= 760 mm Hg at 0°C (millimeters mercury pressure)
	= 29.921 in. Hg at 0°C
	= 14.696 psia
	= 1.01325×10^5 N/m^2
	= 1.01325×10^5 Pa (pascal)
	= 101.325 kPa
1 bar	= 1×10^5 N/m^2
	= 100 kPa
1 psia	= 6.89476×10^3 N/m^2

Energy

1 J (Joule)	= 1 N.m
	= 1 kg.m^2/s^2
	= 10^7 gm.cm^2/s^2 (erg)
	= 9.48×10^{-4} Btu
	= 0.73756 lb$_f$.ft
1 cal (thermochemical)	= 4.1840 J
1 Btu	= 252.16 cal (thermochemical)
	= 778.17 lb$_f$.ft
	= 1.05506 kJ

1 quadrillion Btu	$= 10^{15}$ Btu
	$= 2.93 \times 10^{11}$ kWhr
	$= 172 \times 10^6$ barrels of oil equivalent
	$= 36 \times 10^6$ metric tons of coal equivalent
	$= 0.93 \times 10^{12}$ cubic feet of natural gas equivalent

Power

1 W (watt)	$= 1$ J/s
1 kW	$= 1000$ W
1 hp (horsepower)	$= 0.74570$ kW
1 quadrillion Btu/year	$= 0.471 \times 10^6$ barrels of oil equivalent/day

Viscosity

1 cp (centipoise)	$= 10^{-2}$ gm/cm.s (poise)
	$= 10^{-3}$ kg/m.s
	$= 10^{-3}$ N.s/m^2
	$= 10^{-3}$ Pa.s
	$= 6.7197 \times 10^{-4}$ lb$_m$/ft.s

Appendix III: Overview of Process Economics

Process economics is an essential element of a good design procedure. The objective of this appendix is to provide an overview of basic concepts in cost estimation and economics of chemical processes. For more details, the reader is referred to Humphreys and Wellman (1996), Peters and Timmerhaus (1991), Garrett (1989) and Ulrich (1984).

III.1 Equipment Cost Estimation

A key aspect in most cost estimation studies is the evaluation of the cost of individual pieces of equipment. In addition to quotations from equipment manufacturers, there are several methods for developing cost estimates. For a typical conceptual design, it is normally acceptable to develop an order of magnitude or preliminary estimate whose accuracy is within $+/- 25$ to 50%. The following methods may be employed to develop order of magnitude estimates:

Cost Indices In some cases, the cost of a piece of equipment is available from a previous study, and it is desirable to evaluate its present cost. Because of inflation and other economic changes, it is necessary to correlate equipment cost as a function of time. In this regard, cost indices are useful tools. A cost index is an indicator of how equipment cost varies over time. The ratio of cost indices at two different times provides an estimate for the extent of equipment-cost inflation between these two times. Hence,

$$\text{Present equipment cost}$$
$$= \left(\frac{\textit{Index value at present time}}{\textit{Index value in year n}} \right) \textit{Equipment cost in year n} \quad \text{(III.1)}$$

Various cost indices are published regularly. A commonly used index is the Marshall and Swift (*M&S*) equipment cost index published in the monthly magazine *Chemical Engineering*. For atmospheric pollution control equipment, the Vatavuk cost index may be used (Vatavuk, 1995). It is not recommended to use cost indices if the updating period exceeds ten years.

Cost Charts Equipment cost may be related to one or more basic sizing criteria. Examples of these sizing criteria include diameter for mass exchange trays, diameter and height for packed columns, heat transfer area for heat exchangers, horse power for pumps and compressors, volumetric flow rate for fans and blowers and weight or volume for storage tanks. Extensive compilations of charts correlating equipment cost to sizing criteria have been published in books, journals, and vendor catalogues.

Size Factoring Exponents It is possible to estimate the cost of equipment from the cost of a similar piece of equipment with different size using size factoring or scaling exponents as follows:

$$\text{Cost of equipment of size b} = \left(\frac{Size\ b}{Size\ a}\right)^x \ cost\ of\ equipment\ of\ size\ a \quad \text{(III.2)}$$

where x is the size factoring or scaling exponent. Tabulated values of the exponent for various pieces of equipment are available in literature (e.g., Peters and Timmerhaus, 1991). If the value of x is unavailable, one may assume $x = 0.6$, thus the name the *six-tenth factor rule*.

In order to illustrate the use of cost indices and scaling factors, let us consider the following example. In 1992, the cost of a 50 m^2 shell-and-tube heat exchanger was \$24,000. Estimate the cost of a 100 m^2 shell-and-tube heat exchanger of a similar type and materials of construction in 1996.

Let us assume that the scaling factor for the heat exchanger is 0.6. Hence, in 1992, the cost of the 100 m^2 shell-and-tube heat exchanger can be estimated from Eq. (III.2) to be

$$\left(\frac{100}{50}\right)^{0.6} 24,000 = \$36,400.$$

In order to update the cost of the exchanger, we use the M&S index whose value in 1992 and 1996 was 943 and 1,039, respectively (see *Chemical Engineering* magazine). Therefore, Eq. (III.1) may be used to give the 1996 cost of the 100 m^2 shell-and-tube heat exchanger to be

$$\left(\frac{1,039}{943}\right) 36,400 = \$40,100.$$

III.2 Fixed Capital Investment

The equipment cost described in the previous section typically refers to purchased equipment cost on a *FOB* (*free on board*) basis. This is the cost of the equipment at the manufacturer's loading docks, shipping trucks, rail cars or barges at the vendors fabrication facility. The purchaser still has to pay for equipment freight, installation, insulation, instrumentation, electric work, piping, engineering work and construction. The fixed capital investment is the cost needed to provide the necessary manufacturing and process facilities. It includes the money necessary for process equipment that are installed, equipped with auxiliaries and ready to operate. Hence, fixed capital investment incorporates purchased equipment cost (on a FOB basis) along with freight, installation, insulation, instrumentation, electric work, piping, engineering work and construction. A common method for evaluating fixed capital cost is to multiply the purchased equipment cost by a constant called the *Lang factor*. Typical values of the Lang factor are 3.87, 4.13, and 4.83 for solid, solid-fluid, and fluid processing plants, respectively (Peters and Timmerhaus, 1991).

For instance, in the aforementioned heat exchanger example, the FOB equipment cost was 40,100. If this exchanger is to operate in an oil refinery (primarily a fluid processing plant), the fixed capital investment is $4.83 \times 40,100 = \$193,700$.

Fixed capital investments are characterized by the fact that they have to be replaced after a number of years commonly referred to as service life or useful life period. This replacement is not necessarily due to wear and tear of equipment. Other factors include technological advances that may render the equipment obsolete. Furthermore, over the useful life of the equipment, the plant should plan to recover the capital cost expenditure. In this regard, the notion of depreciation is useful. Depreciation or amortization is an annual allowance which is set aside to account for the wear, tear, and obsolescence of a process such that by the end of the useful life of the process, enough fund is accumulated to replace the process. The simplest method for determining depreciation is referred to as the straight line method in which

$$d = \frac{V_0 - V_s}{n} \tag{III.3}$$

where d is the annual depreciation, V_0 is the initial value of the property, V_s is the salvage value of the property at the end of the service life and n is the service life of the property in years. The limits on annual depreciation are set by federal tax regulations.

The annual depreciation of a fixed capital investment is referred to as *annualized fixed cost*. It allows the company to set aside an annual portion of profit and use it for purposes of capital-cost recovery and tax deduction.

As an illustration, let us revisit the heat exchanger example. The initial fixed cost of the system is $193,700. If the salvage value of the system is negligible and the service life is taken as five years, the annualized fixed cost of the system can be obtained using Eq. (III.3) to be 193,700/5 = $38,700.

III.3 Total Capital Investment

The total capital investment of a process is defined as follows:

$$\text{Total capital investment} = \text{Fixed capital investment}$$

$$+ \text{Working capital investment} \qquad (\text{III.4})$$

where the working capital accounts for the money needed to start up the process, provide initial (e.g., one-month) supply of raw materials, semi-finished and finished products in stock, initial operating expenses and accounts payable/receivable. Typically, the working capital investment is about 15% of total capital investment. For instance, in order to calculate the total capital investment of the heat exchange system, one may use Eq. (III.4):

$$\text{Total capital investment} = 193,700 + 0.15 \text{ Total capital investment},$$

i.e., total capital investment = $227,900.

III.4 Total Annualized Cost

In addition to the fixed capital investment needed to purchase and install process equipment and auxiliaries, there is a continuous expenditure referred to as operating cost, which is needed to operate the process. The operating cost (or manufacturing cost or production cost) includes raw materials, mass-separating agents, utilities (fuel, electricity, steam, water, refrigerants, air, etc.), catalysts, additives, labor, and maintenance. The total annualized cost of a process is defined as follows:

$$\text{Total annualized cost} = \text{Annualized fixed cost}$$

$$+ \text{Annual operating cost} \qquad (\text{III.5})$$

For example, in the aforementioned heat exchanger example let us consider that water is used as a coolant at the rate of 120.0 gallons per minute (gpm) at a cost of $0.20/1000 gallons. Annual labor and maintenance is estimated to be $31,000/year. Hence,

Annual operating cost

$$= 120 \frac{gallons}{minute} \times \frac{\$0.20}{1000 \ gallons} \times \frac{60 \times 8760 \ minutes}{year} + 31,000$$

$$= \$43,600/\text{yr}$$

and the total annualized cost is 38,700 + 43,600 = $82,300/yr.

III.5 Profitability Criteria

The gross income (or profit) of a process is defined as follows:

Annual gross income (or profit) = Annual sales (or savings)

$$- \text{ Annual operating cost} \qquad \text{(III.6)}$$

After federal, state and city taxes are paid, the remaining income is called net income, i.e.,

Annual net income (or profit) = Annual gross income − Annual taxes   (III.7)

There are various indicators to determine the measure of profit for a process. In the following, we describe two of these indicators; return on investment and payout period. The rate of return on investment (ROI) may be calculated as follows:

$$\text{ROI} = \frac{Annual\ net\ income}{Total\ capital\ investment} \times 100\% \qquad \text{(III.8)}$$

The higher the ROI, the more attractive the process. Let us consider the previous heat exchanger system, which costs a total capital investment of $227,900. The heat exchange system will recover via condensation a volatile organic compound whose annual value is $115,000. Because of the environmental nature of the system, the plant was able to get full tax exemption. The plant owner wishes to compare two alternatives; investing in the heat exchange system versus depositing the $227,900 in a savings account in the bank which yields an interest rate of 8%.

According to Eq. (III.6), Annual gross income (in this case equal to annual net income due to tax exemption) = 115,000 − 43,600 = $71,400/yr. Therefore,

$$\text{ROI} = \frac{71,400}{227,900} \times 100\%$$

$$= 31\%$$

which is much higher than the interest provided by the savings account. Therefore, the plant owner should invest in the heat exchange system.

Another profitability criterion is the payback (or payout) period which is defined as follows:

$$\text{Payback period} = \frac{Fixed\ capital\ investment}{Annual\ net\ income + Annual\ depreciation} \qquad \text{(III.9)}$$

It is an indication of how fast the plant can recover the initial fixed capital investment. The shorter the payback period, the more attractive the process. For example, the payback period for the heat exchange example is

$$\text{Payback period} = \frac{193,700}{71,400 + 38,700}$$
$$= 1.8 \text{ years}$$

which is an attractive investment.

References

Garrett, D. E. (1989). Chemical engineering econmics. Van Nostrand Reinhold, New York.

Humphreys, K. K., and Wellman, P. (1996). Basic cost engineering, Third Edition. Marcel Dekker, New York.

Peters, M. S., and Timmerhaus, K. D. (1991). Plant design and economics for chemical engineers, Fourth Edition. McGraw Hill, New York.

Ulrich, G. D. (1984). A guide to chemical engineering process design and economics. John Wiley & Sons, New York.

Vatavuk, W. M. (1995). A Potpourri of equipment Prices. *Chem. Eng.* (August), 68–73.

Appendix IV: Instructions for Software Package

The book is accompanied by a CD-ROM which contains two PC-based software packages; the Mass Exchange Network (MEN) software and LINGO. The following sections summarize the basic features and installation instructions for the two packages.

IV.1 MEN Software

The MEN software is based on the information described in Chapters Five and Six for developing algebraic and optimization-based solutions for the MEN synthesis problem. It can generate composition-interval diagrams, tables of exchangeable loads and optimization formulations for minimizing cost of MSAs.

Installation

Run the SETUP.EXE program from the \MEN\DISK1 directory of the CD and follow the on-screen instructions.

Running the Program

Once you have completed the installation procedure you can run the program by selecting the MEN program icon from the Start Menu (Windows 95) or from the program group (Windows 3.1).

Defining a New Problem

You can start a new problem by selecting *New* from the File menu.

Each MEN problem has lean streams and rich streams. You need to add at least one lean stream and one rich stream before you will be able to generate any meaningful output. To add lean or rich streams, pull down the Streams menu and select either *Add Lean Stream* or *Add Rich Stream*.

For lean streams, you are asked to enter supply composition, target composition, maximum MSA flowrate, solute distribution (m), constant (b), cost, and epsilon (minimum composition difference). If an MSA has no upper limit to flowrate, enter a large number or leave the flowrate as zero in this input field. Essentially all external MSAs fall into this category, while many internal MSAs are only available in certain quantities.

For rich streams you are asked to enter supply composition, target composition, and flowrate.

Continue to add lean and rich streams until you have entered them all.

You can modify stream data by selecting *Edit Lean Stream* or *Edit Rich Stream* from the Streams menu.

You can remove streams from the problem by selecting *Remove Lean Stream* or *Remove Rich Stream* from the Streams menu.

Generating Output

Once you have specified all stream data, you can generate output. There are four options on the Run menu:

- Generate CID (draws the composition interval diagram)
- Generate TEL for Rich Streams (draws Table of Exchangeable Loads for Rich Streams)
- Generate TEL for Lean Streams (draws Table of Exchangeable Loads for Lean Streams)
- Write a LINGO Input File (writes a file to disk for use with LINGO optimization software)

Selecting any of these options causes the appropriate calculations to be performed, and then the results are displayed.

View

Select *View | Input Specifications* from the pull down menu to return to the display of your stream data.

File Options

- *File | New* allows you to reset all variables and start defining a new problem.
- *File | Open* allows you to load a data file (stream data) that you previously saved with this program.

- *File* | *Save* and *File* | *Save As* save stream data to a file, with *Save As* prompting you for a file name.
- *File* | *Print* prints the current window.
- *File* | *Print* Setup allows you to select print options. You will probably want to use this to print some output in portrait printing mode and some in landscape printing mode.
- *File* | *Exit* exits the MEN software.

Special Note on Printing

This program uses TrueType fonts to display information in each window. When you print these windows, you need to make sure that the *Print TrueType fonts as Graphics* option is enabled under the Printer Setup options. Options vary for various printers and their corresponding printer drivers, but most printer drivers either have this option or (less commonly) print TrueType fonts as graphics all the time. If this option is not selected, the printer will probably pick one of its built-in fonts, causing the printout to look different from what is displayed on the screen.

Printing the CID on a Color Printer

Before printing the CID screen on a color printer, hit CTRL-B on the keyboard. This toggles OFF the Force Black/White printing flag. This flag was implemented and turned ON by default. You can toggle between ON and OFF status by hitting CTRL-B at any time.

IV.2 LINGO Software

LINGO is an optimization software that solves linear, nonlinear, and mixed integer linear and nonlinear programs. It is developed by LINDO Systems Inc. The enclosed software is a student version which is limited to solving problems with less than 50 constraints of which up to seven can be nonlinear, and 75 variables of which up to 30 may be integer. Larger versions can be obtained from LINDO Systems Inc. The following is a brief description of how to get started on using LINGO. Additional information can be obtained through the Help menu on the LINGO software or from *LINGO User's Guide* published by LINDO Systems Inc., Chicago (1995).

Installation

Run the SETUP.EXE program from the \LINGO directory of the CD and follow the on-screen instructions.

Running the Program

Once you have completed the installation procedure you can run the program by double clicking on the LINGO.EXE program icon.

Creating a New File

From the File menu, use the *New* command. An empty "untitled" window will be displayed. On this window, you can type the model.

Model Format

The following is some basic information on writing a LINGO optimization model. The general form of the model is as follows:

MODEL:
Min (or Max) = Objective function;
Constraints, each followed by;
END

If the line for minimizing or maximizing an objective function is not included, LINGO will solve the model as a set of equations provided that the degrees of freedom are appropriate. In writing constraints, the equalities and inequalities can be described as follows:

=   The expression to the left must equal the one on the right.

<= The expression to the left must be less than or equal to the expression on the right.

>= The expression to the left must be greater than or equal to the expression on the right.

<   The expression to the left must be strictly less than the expression on the right.

>   The expression to the left must be strictly greater than the expression on the right.

In writing constraints, the following symbols are used for mathematical operations:

+   Addition
−   Subtraction
∗   Multiplication
/   Division
∧   Power

For example, let us consider the following simple minimization program:

Minimize x + y

Subject to the following constraints

$$x \geq 5.5$$

$$y \geq 7.0$$

In terms of LINGO modeling language, the program can be written as follows:

MODEL:

Min = x + y;

X >= 5.5;

Y >= 7.0;

END

Solving a Program

From the LINGO menu, choose solve. LINGO will display a LINGO Solver Status window and a Reports window that shows the following solution to the previous program:

Objective value: 12.50000

Variable   Value

X   5.500000

Y   7.000000

It indicates that the minimum value of the objective function is 12.5 and that the optimal solution for x and y is 5.5 and 7.0, respectively.

Mathematical Functions

The following are some of the mathematical functions used by LINGO:

@ABS(X)   Returns the absolute value of X.
@EXP(X)   Returns the constant e (2.718281...) to the power X.
@LOG(X)   Returns the natural logarithm of X.
@SIGN(X)   Returns -1 if X is less than 0, returns $+1$ if X is greater than or equal to 0.
@BND(L, X, U) Limits the variable or attribute X to greater or equal to L and less than or equal to U.
@BIN(X)   Limits the variable or attribute X to a binary integer value (0 or 1.)
@GIN(X)   Limits the variable or attribute X to only integer values.

For example, if x in the previous program is to assume only an integer value, the program should be extended to:

MODEL:

Min $= x + y$;

X $>= 5.5$;

Y $>= 7.0$;

@GIN(X);

END

By using the Solve option from the LINGO menu, we get the following results displayed on the Reports window:

Objective value: 13.00000

Variable Value

X 6.000000

Y 7.000000

Index

A

Acrylonitrile, 86
Allocation of species, 11
Ammonia, 40, 87, 235, 260
Ammonium nitrate, 240

B

Benzene, 33, 38, 53, 135, 150
Bhopal, 289

C

Carbaryl, 289
Carbon dioxide, 123
Cascade diagram for mass exchange, 107
Cascade diagram for heat exchange, 225
CHARMEN (*see* Combined heat and reactive mass exchange network)
Chloroethanol, 7–10
CID (*see* Composition interval diagram),
Coal liquefaction, 207
Cold composite stream, 220
Coke oven gas, 75
Combined heat and reactive, mass exchange network, 217, 232
Composition interval diagram, 105, 106, 126
Compression energy, 29
Continuous mass exchangers, 23–26
Copper etching, 148
Corresponding composition scales, 47, 49, 193, 195

D

Debottlenecking, 86, 94
Desulfurization, 75, 123, 195, 207, 209, 211, 242
Disposal, 2

E

EARs (*see* Environmentally acceptable reactions)
Economic potential, 151
Efficiency of mass exchange, 23
EISENs (*see* Energy induced separation networks)
End-of-pipe treatment, 1, 2
Energy induced separation networks, 259
Energy integration, definition of, 10
Energy separating agents, 12
Environmentally acceptable reactions, 288
Equilibrium, 17–19
ESAs (*see* Energy separating agents)
Ethyl benzene, 77, 149
Ethyl chloride, 7–10
Ethylene, 77
Excess capacity of MSAs, 52
External MSAs
  cost estimation, 64
  cost minimization, 68
  screening, 68, 69

F

Feasibility criteria for mass exchangers, 112–114